W0263325

Recent Results in Cancer Research
Fortschritte der Krebsforschung
Progrès dans les recherches sur le cancer

II

Edited by

V. G. Allfrey, New York · M. Allgöwer, Chur · K. H. Bauer, Heidelberg · I. Berenblum, Rehovoth · F. Bergel, London · J. Bernard, Paris · W. Bernhard, Villejuif N. N. Blokhin, Moskva · H. E. Bock, Tübingen · P. Bucalossi, Milano · A. V. Chaklin, Moskva · M. Chorazy, Gliwice · G. J. Cunningham, London · W. Dameshek, Boston M. Dargent, Lyon · G. Della Porta, Milano · P. Denoix, Villejuif · R. Dulbecco, San Diego · H. Eagle, New York · R. Eker, Oslo · P. Grabar, Paris · H. Hamperl, Bonn · R. J. C. Harris, London · E. Hecker, Heidelberg · R. Herbeuval, Nancy · W. C. Hueper, Bethesda · H. Isliker, Lausanne · D. A. Karnofsky, New York · J. Kieler, København · G. Klein, Stockholm · L. G. Koss, New York G. Martz, Zürich · G. Mathé, Paris · O. Mühlbock, Amsterdam · G. T. Pack, New York · V. R. Potter, Madison · L. Sachs, Rehovoth · E. A. Saxén, Helsinki W. Szybalski, Madison · H. Tagnon, Bruxelles · R. M. Taylor, Toronto · A. Tissières, Genève · E. Uehlinger, Zürich · T. Yoshida, Tokyo · L. A. Zilber, Moskva

Editor in chief

P. Rentchnick, Genève

Springer-Verlag Berlin Heidelberg New York 1966

Neuroblastomas
Biochemical Studies

Edited by

C. Bohuon

With 29 Figures

Springer-Verlag Berlin Heidelberg New York 1966

Colloque sur la Biochimie des Neuroblastomes
Institut Gustave Roussy, Villejuif
29 et 30 mai 1964

Sponsored by the Swiss League against Cancer

ISBN-13: 978-3-642-94972-2   *e-ISBN: 978-3-642-94972-2*
DOI: 10.1007/978-3-642-94972-2

Introduction

Les différentes communications contenues dans cette monographie ont été présentées lors d'un Colloque sur la Biochimie des Neuroblastomes organisé à l'Institut Gustave Roussy les 29 et 30 Mai 1964. Il m'avait semblé, et j'avais été encouragé en ce sens par le Dr. SCHWEISGUTH et le Dr. LA BROSSE, qu'il était temps que les spécialistes de cette question puissent se rencontrer pour exposer leurs travaux, échanger des idées, organiser les recherches si possible à l'échelon international. Le Professeur agrégé DENOIX, Directeur de l'I.G.R. a accueilli très favorablement ce projet et a facilité considérablement l'organisation de ces journées.

Depuis quelques années, en effet, des publications montraient l'intérêt suscité chez les cancérologues et les pédiatres par cette tumeur secrétante de l'enfant, rare, mais non exceptionnelle. Il était nécessaire après quelques années, de faire un premier bilan des acquisitions. Ce Colloque a permis à la plupart des spécialistes mondiaux de «faire connaissance». Comme les lecteurs pourront s'en rendre compte, les communications présentées ont toutes été d'une très grande qualité.

Je remercie tous ceux qui ont bien voulu participer à ce Colloque; j'espère que nous nous retrouverons encore plus nombreux lors du prochain Colloque prévu dans le courant de 1966.

Professeur agrégé C. BOHUON

Contents

Quantitative Determination of Metanephrine and Normetanephrine in Urine. By KAZUMI TANIGUCHI, YASUO KAKIMOTO and MARVIN D. ARMSTRONG. With 5 Figures 1

Réflexions sur le Dosage de la Catéchol O-Méthyl Transférase et de la Mono-amine Oxydase dans les Neuroblastomes. Par C. BOHUON, E. H. LA BROSSE, M. ASSICOT et A. AMAR-COSTESEC 16

Problèmes posés à l'Organicien par la Synthèse des Métabolites non Azotés des Catécholamines. Par JEAN GARDENT et JOSEPH LIKFORMAN 22

Cystathioninuria and Vanil-lactic-acid-uria. By LEIV R. GJESSING. With 2 Figures 26

Mechanism of Action of α-Methyl-Dopa. By H. KÄSER and H. P. WAGNER. With 3 Figures 33

Biochemical Features of Malignant Tumours of Sympathetic Tissues. By RONALD ROBINSON. With 1 Figure 37

Observations on the Biochemical Diagnosis of Neuroblastoma. By MAURICE BELL. With 4 Figures 42

Formation et Renouvellement de l'Acide 3-Méthoxy-4-hydroxymandélique dans les Neuroblastomes. Par E. H. LA BROSSE. Avec 4 Figures 51

4-Hydroxy-3-Methoxyphenylglycol and other Compounds in Neuroblastoma. By M. SANDLER and C. R. J. RUTHVEN. With 2 Figures 55

Intérêt pour le Pédiatre des Dosages d'Acide Vanillylmandélique dans les Urines. Par O. SCHWEISGUTH. Avec 7 Figures 59

Enzymes Controlling the Biosynthesis of Catecholamines. By M. GOLDSTEIN. With 1 Figure 66

Conclusion 71

List of Authors

AMAR, A., Dr., Unité de Biologie expérimentale, Institut Gustave Roussy, Villejuif (Seine) France.

ARMSTRONG, MARVIN D., Professeur, Department of Biochemistry The Fels Research Institute, Yellow Springs (Ohio), USA.

ASSICOT, M., Dr., Unité de Biologie expérimentale, Institut Gustave Roussy, Villejuif (Seine) France.

BELL, MAURICE, Professor Dr., Biochemistry Department, Manchester Royal Infirmary, Manchester/Great Britain.

BOHUON, C., Professeur Agrégé, Unité de Biologie expérimentale, Institut Gustave Roussy, Villejuif (Seine) France.

LA BROSSE, E. H., Dr., Shock Trauma Unit. Depart. of Surgery, University of Maryland School of medicine, Baltimore I (Maryland 21 201) USA.

GARDENT, JEAN, Professeur Agrégé, Pharmacie Centrale des Hôpitaux de Paris, Paris/France.

GJESSING, LEIV R., Dr., Central Laboratory, Dikemark Hospital, Asker/Norway.

GOLDSTEIN, M., Professor, Department of Biochemistry, N. Y. University School of Medicine, New York, USA.

KÄSER, H., Dr., Swiss Center of Clinical Tumor Investigation, Institute for Clinical Protein Research and Pediatric Clinic, University of Berne, Berne/Switzerland.

KAKIMOTO, YASUO, Dr., Department of Biochemistry, The Fels Research Institute, Yellow Springs (Ohio), USA.

LIKFORMAN, J., Dr., Pharmacie Centrale des Hôpitaux de Paris, Paris France.

ROBINSON, RONALD, Ph. D., Group Pathological Laboratory, Warwick/Great Britain.

RUTHVEN, C. R. J., Dr., Bernhard Baron Memorial Research Laboratories, Queen Charlotte's Maternity Hospital, London/Great Britain.

SANDLER, M., Dr., Bernhard Baron Memorial Research Laboratories, Queen Charlotte's Maternity Hospital, London/Great Britain.

SCHWEISGUTH, MLLE. O., Dr., Unité de Biologie expérimentale, Institute Gustave Roussy, Villejuif (Seine) France.

TANIGUCHI, KAZUMI, Dr., Department of Biochemistry, The Fels Research Institute, Yellow Springs (Ohio), USA.

WAGNER, H. P., Dr., Swiss Center of Clinical Tumor Investigation, Institute for Clinical Protein Research and Pediatric Clinic, University of Berne, Berne/Switzerland.

The Fels Research Institute, Yellow Springs, Ohio

Quantitative Determination of Metanephrine and Normetanephrine in Urine [1]

By

Kazumi Taniguchi, Yasuo Kakimoto, and Marvin D. Armstrong

With 5 Figures

Metanephrine and Normetanephrine in Urine

I. Abstract

Details of a fluorometric method for the separate determination of free and conjugated metanephrine and normetanephrine in urine are presented. The procedure consists of: (1) a separation and concentration of free amines by absorption on a column of Dowex-50 resin followed by elution with a small volume of aqueous ammonia; (2) acid hydrolysis of the initial effluent from the Dowex-50 column to release conjugated amines, which are then separated in the same manner as for free amines; (3) separation of the metanephrine and normetanephrine in each fraction by chromatography on a column of Amberlite CG-50; and (4) a fluorometric estimation of the amines by a variation of the trihydroxyindole method. Metanephrine may be estimated in samples which contain between 0.25 and 50 μg. and normetanephrine between 0.35 and 120 μg. No substances have been encountered in urine which interfere with fluorescence development or which respond to the test. The average daily excretion of the amines by 6 male subjects was: metanephrine, free, 29.5 μg., conjugated, 134.1 μg.; total, 163.6 μg.; and normetanephrine, free, 21.2 μg.; conjugated, 223.2 μg.; and total 244.4 μg. The amounts of the different fractions excreted by an individual appear to be relatively constant, and marked differences in the pattern of excretion was observed in different persons. There were marked diurnal variations in the rate of excretion of the amines but there was no consistent pattern to the variation.

In recent years the diagnosis of pheochromocytoma and other tumors of the sympathetic nervous system has become facilitated by the development of methods which are capable of detecting the characteristically increased excretion of epinephrine and norepinephrine and their metabolites. The catecholamines may be

---

[1] This work was supported in part by Research Grant MH-2278 from the National Institute of Mental Health, U.S., Public Health Service.

detected readily with the fluorometric method of VON EULER and FLODING [1], and various procedures have been reported for the determination of their 3-O-methyl derivatives, metanephrine and normetanephrine [1] [2], and of the major end product of the metabolism of all the amines, 3-methoxy-4-hydroxy-mandelic acid [3]. At the present time there is considerable interest in the excretion of catecholamines and their O-methyl derivatives by persons who do not have neural tumors. Variations in the small amounts excreted by the same person and between different individuals are of interest in connection with research on psychiatric disorders and also because of a possible correlation with normal differences in psychological constitution.

There should be one major advantage attached to the use of measurements of M and NM in urine rather than of epinephrine and norepinephrine. The catecholamines are unstable to handling in neutral or slightly alkaline solutions, so that it has been difficult to devise routine procedures for separating them. In addition, precautions are necessary to prevent their destruction during the collection and storage of samples. The O-methylated amines are much more stable chemically, and may be handled with comparative safety. Several methods for measuring urinary M and NM have been reported recently [4—7], but each of these suffers from some disadvantage. YOSHINAGA et al. [4], extracted amines from hydrolyzed urine which had been adjusted to pH 10, subjected the amines to high-voltage electrophoresis, eluted the dyes produced by coupling them with diazotized p-nitroaniline, and measured the dyes spectrophotometrically. They did not report separate values for free and conjugated amines, the extraction procedure is somewhat laborious and subject to errors, and apparatus for high voltage electrophoresis is not standard equipment in all laboratories. KRAUPP et. al [5] also extracted amines from hydrolyzed urine, subjected them to twodimensional paper chromatography, and eluted dyes formed from them and measured them spectrophotometrically. SMITH and WEIL-MALHERBE [6] hydrolyzed urine, removed interfering materials by electrodialysis, separated the catecholamines by adsorption onto alumina, and measured M and NM by differential fluorometry. With this procedure free and conjugated amines would have to be measured by using parallel determinations and subtracting the free from total amines to estimate the conjugated fraction. The use of differential methods, particularly for two phases of the determinations, has inherent disadvantages. BRUNJES et al. [7] used an ion-exchange resin to separate amines from hydrolyzed urine and determined M an NM by differential fluorometry. Their procedure has the same disadvantage as that of SMITH and WEIL-MALHERBE, though the steps involving the electrodialysis procedure and removal of the interfering catecholamines are avoided.

Observations made during work on characterizing the phenolic amines in urine [8] offered encouragement that a simple procedure might be devised for determining M and NM. First, free amines were found to be quantitatively held on sulfonated ion exchange resins while conjugated amines passed through columns of the resins. Next, M and NM were found to be separated readily by chromatography on ion-exchange resins. Finally, in the course of characterizing the amines in urine it was observed that M and NM yielded intensely fluorescing areas on

---

[1] Abbreviations used: metanephrine, M; normetanephrine, NM.

chromatograms which had been sprayed with periodate reagent [8]. These observations were extended and led to the method reported below.

II. Reagents

1. 0.1 N sodium acetate.
2. 5 N ammonium hydroxide.
3. Glacial acetic acid.
4. 5 N sodium hydroxide.
5. 0.1% potassium meta periodate (KIO_4) (Matheson Coleman and Bell, Reagent grade).
6. Borate buffer, pH 8.70. Dissolve 778 g. of boric acid (H_3BO_3) (Matheson Coleman and Bell, A. R., granular) in water and adjust the final volume to exactly 25 l. Add about 8.7 l. of 0.5 N sodium hydroxide to adjust the pH to as nearly 8.70 as possible. A large volume of this buffer should be prepared because each batch needs to be checked as described later to establish conditions for the separation of metanephrine and normetanephrine.
7. Standard metanephrine solution. Dissolve an amount of metanephrine hydrochloride (Calbiochem) in a mixture of borate buffer, pH 8.70, and glacial acetic acid (50 : 1) to provide a concentration of 10 μg./ml. of the free base.
8. Standard normetanephrine solution. Dissolve an amount of normetanephrine hydrochloride (Calbiochem) in a mixture of borate buffer, pH 8.70, and glacial acetic acid (100 : 1) to provide a concentration of 10 μg./ml. of the free base. The standard solutions should be kept in a refrigerator, where they are stable for several months.
9. 0.2% ascorbic acid in 5 N sodium hydroxide. This reagent must be prepared fresh and used immediately. While the periodate oxidation is being effected, as described later, 30 mg. of ascorbic acid (Merck, U. S. P., fine crystals) is weighed into a 50 ml Erlenmeyer flask. At the end of the oxidation 15 ml. of 5 N sodium hydroxide is added to the flask, which is swirled to dissolve the crystals, and the solution is pipetted into the reaction tubes immediately.

III. Preparation of Resins and Columns

1. Dowex 50 X 2 (H^+), 100—200 mesh. A pound of resin is placed in a 4 l. beaker, 2 l. distilled water is added, the mixture is stirred and the resin is allowed to soak overnight. The mixture is stirred again, the resin is allowed to settle, and fines are decanted with the water. More water is added and the settling and decanting are repeated until most of the fines have been removed. A slurry of the resin is poured into a large column (7 × 90 cm.) and the resin is washed successively with 1 l. of 4 N hydrochloric acid, with water until the pH of the effluent is about 5, with 2 l. of 2 N sodium hydroxide, with water until the pH is less than 6, and with 1 l. of 4 N hydrochloric acid to convert the resin to the H^+ form. It is then washed with water until the pH of the effluent is about 5, removed from the column, and sored with an equal volume of water. Resin which is to be stored for

an extended time should be kept under water in a refrigerator. A suspension of resin is poured into a chromatographic column to a packed resin volume of 4 ml.

Chromatographic columns 10×300 mm. (Corning 38 450 or similar) are convenient for the concentration and separation of the amines. The bottoms of the columns are stoppered, volumes of 4 ml. and 12 ml. of water are added, and the columns are permanently marked for these volumes; the height of the resin column is about 5 cm. for the 4 ml. volume and 15—16 cm. for 12 ml. For convenience in running several analyses simultaneously a wooden or metal rack can be made to hold the columns; eight columns are the most that can be operated comfortably by one person. A convenient device for adding samples and larger volumes of eluents to the columns provided by sealing a 6—8 cm. length of 8 mm. tubing to the bottom of a 125 ml. Erlenmeyer flask. A small section of rubber tubing is slipped over the glass tubing near the base to provide an airtight seal when the tubulation is inserted in the chromatographic tube which supports the flask.

2. Amberlite CG-50, 200—400 mesh. A pound of resin is soaked in water and fines are removed in the manner described above. The resin is then treated with $4 N$ hydrochloric acid, water, $2 N$ sodium hydroxide, and water in the beaker, with successive decantations. The resin is then packed into a large column and washed again with $4 N$ acid, water, $2 N$ alkali, and water. Finally, the resin is washed with the borate buffer, pH 8.70, until the pH of effluent is the same as the buffer. The resin is removed from the column and stored under buffer. For use, a suspension is poured into a 10×300 mm. column until the volume of packed resin is exactly 12 ml.

IV. Procedure

1. Concentration of the Amines from the Urine

a) Free amines. The urine sample is filtered to remove any sediments present and the creatinine content is determined with the Jaffé method. An amount of urine which contains 100 mg. of creatinine is diluted with water to a volume of 200 ml. and the pH is adjusted to between 4 and 6 by the addition of either acetic acid or ammonium hydroxide as necessary. Urine with a creatinine content less than 0.5 mg. creatinine per ml. may be used directly, but more concentrated samples should be diluted. The diluted urine is allowed to flow through the 1×5 cm. (4 ml. volume) column of Dowex 50×2 (H^+) by gravity. The resin is washed with 10 ml. of water and the combined effluents are saved for the estimation of conjugated amines. The resin is then washed with 20 ml. of 0,1 M sodium acetate and 10 ml. of water successively; this eliminates amino acids and other substances which interfere with the separation of M and NM. After the washings, the amines which were held on the resin are eluted by the passage of 8 ml. of 5 N ammonium hydroxide through the column. The eluate is dried either by leaving it in a vacuum desiccator overnight over sulfuric acid or by evaporation of the water under reduced pressure. In the latter case, care should be taken that excessive foaming does not occur while ammonia is removed and the temperature should not exceed 40°. The resin is discarded after it has been used once.

b) Conjugated amines. To one-half of the combined effluents (50 mg. creatinine equivalent) saved above, concentrated hydrochloric acid is added until the final acid concentration is 0.1 N; the solution is then heated on a steam bath for thirty minutes. After hydrolysis is completed, the sample is cooled to room temperature and the liberated amines are separated in the same manner as described for the free amines, with the exception that the first water wash is omitted. The resin takes up the amines effectively from the acid solution.

c) Direct hydrolysis of urine. If separate determination of free and conjugated amines is not required or if the available urine sample is not sufficient for their estimation, an amount of urine which contains 30 to 50 mg. of creatinine is diluted to contain 0.5 mg. creatinine per ml. and treated in the manner described for hydrolyzing and separating the conjugated amines.

2. Separation of Metanephrine and Normetanephrine

The dried material containing the amines is dissolved in 2 ml. of pH 8.70 borate buffer, transferred quantitatively to the column of Amberlite CG-50 with a capillary pipette, and the container is washed with an additional 1 ml. of the buffer, which is also applied to the column. The amines are developed and eluted with the same buffer. The eluate from the column is collected in graduated cylinders. The first 36 ml. of effluent is discarded; the next 25 ml. fraction, which contains M (Fraction 1), and the following 33 ml. fraction, which contains NM (Fraction 2), are collected separately. Short lengths of rubber tubing are placed on the tips of the columns, and pinch clamps may be used to control the flow of the buffer. After the sample has been applied to the resin and the flow of eluting buffer has started the flow may be stopped at any time without disturbing the chromatography of the amines.

3. Determination of Metanephrine and Normetanephrine

a) Metanephrine. To fraction 1, 0.50 ml. of glacial acetic acid is added to adjust the pH to 4.5, and three 3 ml. portions of the solution are placed in three test tubes. To the first tube, 0.2 ml. of 5 N NaOH is added; this tube serves as a reagent "blank". To the second tube, 0.2 ml. of a mixture of acetic acid and the borate buffer (1 : 50) is added. To the third tube, 0.2 ml. of the standard solution of M is added; this serves as an "internal standard". A half ml. of 0.1% potassium meta-periodate solution is added to each tube and the solution is allowed to stand at room temperature for forty minutes. One ml. of freshly prepared solution of 0.2% ascorbic acid in 5 N sodium hydroxide is then added rapidly to each tube with shaking.

b) Normetanephrine. Fraction 2 is acidified to pH 5 by the addition of 0.36 ml. of glacial acetic acid, and three 3 ml. portions are used for the development of fluorescence from NM in the same manner as described above for metanephrine, with the exception that a 1 : 100 mixture of glacial acetic acid and the borate buffer is used to dissolve the standard compound and to adjust the volume of the second tube.

Three minutes after the addition of alkaline ascorbic acid solution, the fluorescence derived from the amines is measured with an Aminco-Bowman spectrophotofluorometer. The wavelengths for activation and fluorescence of the fluorophores are 415 and 520 mμ., respectively, for metanephrine, and 405 and 510 mμ., respectively, for normetanephrine. The fluorescence measurements should be completed within ten minutes after the addition of the alkaline ascorbic acid solution. With the Aminco-Bowman spectrophotofluorometer the following size and placements of slits were used: 1, none; 2, 1/16 in.; 3 and 4, 3/16 in.; 5, 1/64 in.; 6, none; and 7, 1/64 in. This arrangement allows a good intensity of activation energy and good resolution for fluorescence measurement. With a sensitivity setting of 50, a meter multiplier setting of 0.1 was used for the blank and sample, and 0.1 for the internal standard. With these conditions, a meter reading of 1—2 units was obtained for the blank for free amines and 2—3 for conjugated amines. Readings for free M in usual samples ranged from 6—30, for conjugated M from 20—90, for free NM 4—9, and for conjugated NM 15—50. The internal standards for M and NM gave readings of 24 and 16 units, respectively.

The following equation is used to calculate the amount (μg.) of amines in the urine sample:

$$\mu\text{g.} = \frac{A}{B} \times \frac{2}{3} \times \frac{\text{Vol. of eluate} + \text{vol. of added acetic acid}}{\text{recovery rate}}$$

Where A = meter reading of sample — meter reading of blank
and   B = 10 $\times$ meter reading of standard — meter reading of sample.
In practice, all the values except A and B remain constant, and are reduced to a factor for use. Representative factors obtained in this work were as follows: M, free, 21.8, conjugated, 18.9; NM, free, 31.4, conjugated, 28.5.

Photofluorometers other than the Aminco-Bowman spectrophotofluorometer could be used with appropriate filters.

V. Discussion

1. Concentration of the Amines by Dowex 50 X 2

Because of the small amounts present in urine, amines must be concentrated and separated from the major components before they can be subjected to chromatography on Amberlite CG-50. Dowex 50 X 2 (H$^+$) is satisfactory for this purpose because of its high capacity for the amines. However, if an amount of urine which contains 100 mg. of creatinine is used for separation of free amines or 50 mg. of creatinine for conjugated amines, the resin also holds other materials which may disturb the elution pattern of the amines on Amberlite CG-50. Many of these substances are eliminated by the 0.1 M sodium acetate wash, even though this washing results in some loss of amines; this loss will be discussed later. Other treatments for preliminary purification of urine samples were attempted, but none could be devised which did not result in a major loss of the amines.

2. Chromatography on Amberlite CG-50

Columns are prepared by pouring a thin suspension of the resin in the borate buffer into the column, with the outlet left open, until the top of the packed resin

reaches the premarked line for a volume of 12 ml. The height of the bed becomes
a little lower after repeated uses, but it is not necessary to make any adjustment
in order to keep the same elution pattern. Columns may be used indefinitely or
until they have packed so much that flow rate of the buffer through them has
become so slow as to be inconvenient. Columns prepared from the same batch of
resin and of borate buffer always showed identical elution patterns for the two
amines, even after they had been used repeatedly. Flow rate of the buffer with
gravity is about 30 ml./hr. with all the columns. After columns have been packed,
each should be washed with 100 ml. of buffer and one of them should be examined
to establish the elution pattern of the amines. This is done by applying a sample
which contains 10 μg. of M and 10 μg. of NM and eluting with the buffer; 3 ml.
fractions should be collected, preferably with a fraction collector if one is available,
and the amine content of each fraction determined either by measuring fluorescence
at 325 mμ. stimulated by activation at 280 mμ. or by fluorophore formation. A
typical pattern is shown in Fig. 1; with this determination it is possible to select

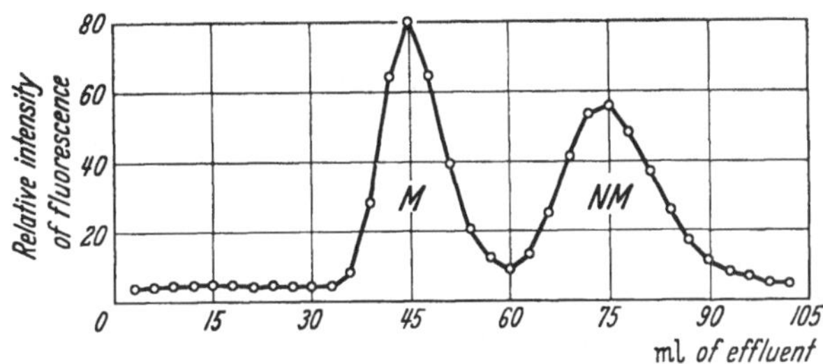

Fig. 1. Elution pattern of metanephrine (M) and normetanephrine (NM) from 1 $\times$ 15 cm. column of Amberlite
CG-50 with borate buffer, pH 8.70. Fluorescence at 325 mμ by activation at 280 mμ

the volumes to be collected for the determinations of M and NM. The elution
pattern varies slightly when different batches of resin or buffer are used. On many
occasions each of the eight columns which were usually used were checked indi-
vidually in this manner and were found to have almost exactly the same properties.
Thus it is usually necessary to test only one column with a given batch of resin
or buffer. However, in a laboratory beginning to use this procedure it is probably
desirable that all the columns be tested to insure accuracy and to assure confidence
in the method. The elution pattern of the amines varies markedly if there are
large changes in room temperature. In this laboratory, room temperature was
24° ± 1° C during all of the experiments. In laboratories without temperature control
where room temperature might vary between 15 and 35° C, it would be necessary
to use jacketed columns with circulating water at constant temperature.

The elution pattern of the amines in the urine concentrates coincides closely
with that of pure amines. However, if the step in which interfering substances are
washed off the Dowex-50 column with 0.1 M sodium acetate is omitted, as discussed
later with the recovery experiments, the urine amines are eluted about 3 ml. earlier
than with the pure compounds. In this case special precautions should be made
in collecting the effluent. In practice, however, slight variations in the elution pattern
of the amines would not lead to a significant error in their measurements.

The column ist ready for reuse as soon as the NM in a sample has been eluted.
Pigments are not held on the resin and are washed out before metanephrine, and

no interfering substance has been encountered in the amine fractions which disturbs the elution pattern or the development of fluorescence. Epinephrine and norepinephrine are destroyed completely when the amines are eluted from Dowex-50 with ammonia, as discussed later. Phenolic amines which may be present and are eluted near metanephrine and normetanephrine do not interfere nor do any of the other amines which might remain on the column and be eluted during subsequent separations of M and NM.

The recommended volume of the resin is optimal for the practical separation of the two amines. If a smaller volume is used, the volumes of eluate which must be collected become critical. If a larger volume is used, too much time is required and the amines are eluted in a larger volume of buffer, which causes a loss of sensitivity.

3. Formation of the Fluorophore

a) **pH optimum for the oxidation.** The effect of pH of the medium on the development of the fluorophores M and NM is shown in Fig. 2. It is apparent that the optimal pH is 4.5 for M and 5.0 for NM. In buffers other than borate the optimal pH for metanephrine oxidation is lower.

b) **Selection of oxidant.** Under the condition used potassium metaperiodate (KIO_4) is effective for the conversion of the amines to fluorophores. With the same conditions iodine (0.1 ml. of 1% solution) [9, 10] and potassium ferricyanide (0.5 ml. of 0.1% solution) [6, 7] did not form any fluorophore. The use of higher concentrations of potassium periodate than 0.1 per cent led to a rapid decrease of the fluorescence after the addition of alkaline ascorbic acid, and a lower concentration resulted in incomplete oxidation.

c) **Time required for oxidation and stability of the fluorophore.** In Fig. 3 the effect on the fluorophore intensity of the time allowed for the oxidation by periodate is shown. Maximum fluorescence developed in 30 minutes and remained stable another 20 minutes at room temperature (24°). After the addition of the alkaline

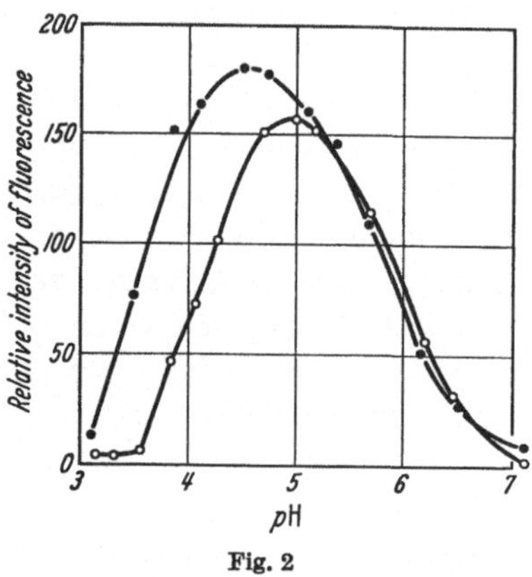

Fig. 2

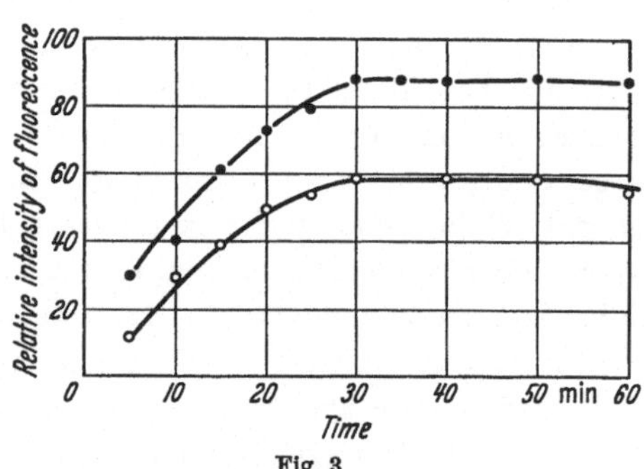

Fig. 3

Fig. 2. Effect of pH on the development of the fluorophores from metanephrine (closed circles) and normetanephrine (open circles). The pH of the reaction mixtures were as indicated in the figure; glacial acetic acid was added to the borate buffer to adjust the pH. Fluorescence was developed and measured as described in the text

Fig. 3. Time course of development of fluorescence by periodate oxidation. Fluorescence was developed and measured as described in the text. Closed circles, metanephrine; open circles, normetanephrine

ascorbic acid the fluorescence becomes maximal after 3 minutes, and is stable for another 10 minutes. Practically, because of the limited stability of the fluorophore after the alkaline ascorbic acid solution has been added, only 12 tubes may be developed in one batch. The concentration of ascorbic acid in 5 N sodium hydroxide required for optimal development of fluorescence was examined and it was found that 0.2 per cent ascorbic acid in 5 N sodium hydroxide is optimal for both amines. Minor variations from this concentration do not result in significant changes in the development of fluorescence.

d) **Use of a blank and an internal standard.** A sample blank is used to guard against the possible presence of some interfering substance in the sample or reagents which might be present and fluoresce without periodate oxidation. This is done most simply by making an aliquot of the sample alkaline to pH 10 by adding 0.2 ml. of 5 N sodium hydroxide. The fluorophores are not formed from the amines at this pH. The same blank value is obtained by reversing the order of addition of alkaline ascorbic acid and periodate or by omitting ascorbic acid from the alkaline ascorbic acid solution.

To guard against the presence of some substance in the sample or reagents which might inhibit or quench development of fluorescence from the amines, 2 μg. of authentic amines are routinely added to samples to provide an internal standard. This also serves to protect against errors which might occur because of fluctuations in the intensity of the exciting light. However, in a large number of determinations made in this laboratory, no interfering substance has been observed in either blank or standard measurements. Therefore, it is usually necessary to use only one blank tube and one standard tube with each batch of samples.

e) **Sensitivity and linearity of the reaction.** If a meter reading double the one obtained for the blank is considered to be the smallest significant one, the minimal

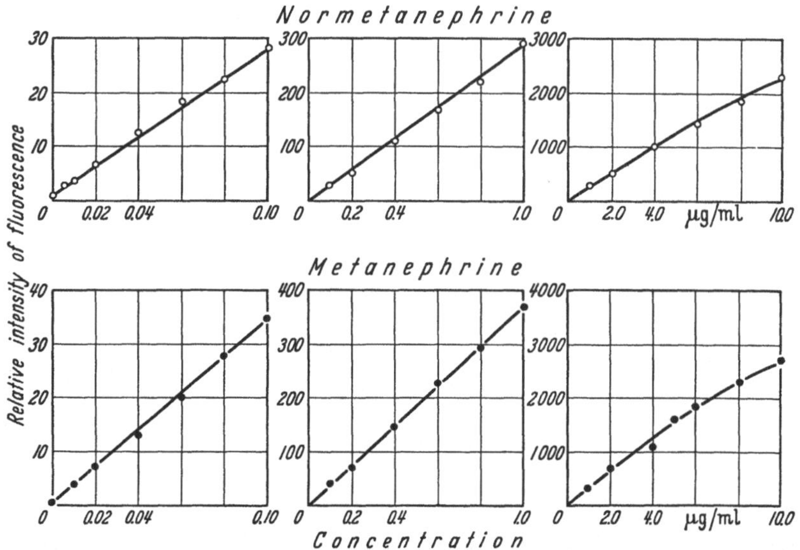

Fig. 4. Sensitivity and linearity of fluorescence. Three ml. samples of the amines at the concentration shown were prepared from their standard solutions in the acidified borate buffer, which was used for dilutions. Fluorescence was developed and measured as described in the text

amount of amine which can be measured in a urine sample is 0.25 μg. of M and 0.35 μg. of NM. Practically, an amount of urine which contains 100 mg. of creatinine is required for the accurate estimation of free amines and 30 to 50 mg. of creatinine for the conjugated or total amines.

The data presented in Fig. 4 show that the intensity of fluorescence is linear up to a concentration of 2.0 μg./ml. of M and 4.0 μg./ml. of NM in the sample used for development of fluorescence. Accurate estimations may be made with urine samples which contain up to about 50 μg. of M and 120 μg. of NM.

f) Specificity of the fluorescence development. Several materials which were known to be retained in the amine fraction and some other substances related in structure to the catechol and guaiacol amines were tested for their ability to form fluorophores under the conditions used here. The results of these experiments are listed in Table 1. The catecholamines and dopa give rise to a strong fluorescence but they are destroyed by the elution of amines from Dowex-50 with ammonia. Although the red pigments formed from the catechols develop fluorescence with the procedure they are not held on the Amberlite CG-50, but are completely washed off before M. 3-Methoxytyrosine is not held well on Dowex-50 and not at all on the Amberlite CG-50 at pH 8.70, so it does not interfere. Serotonin and 3-methoxytyramine give rise to a weak fluorescence, but both are eluted from the column

Table 1. *Specificity of fluorescence development*

Solutions of 0.02 μ mole of each of the substances listed below in 3 ml. of borate buffer adjusted to pH 5.0 were subjected to the conditions described in the text for the development of fluorescence.

Substance	Relative intensity of fluorescence	
	Metanephrine 415 mμ activation 520 mμ fluorescence	*Normetanephrine* 405 mμ activation 510 mμ fluorescence
Metanephrine	420	370
Normetanephrine	262	330
Epinephrine	605	520
Norepinephrine	336	425
Dopamine	4	6
Dopa	5	18 (155)*
3-Methoxytyrosine	2.5	22 (130)*
3-Methoxytyramine	0.5	1
Serotonin	2	1
Adrenalone	1	1.5
p-Sympatol	1	1
N-Methylmetanephrine	0	0
Octopamine	0	0
Tryptamine	0	0
p-Tyramine	0	0
m-Tyramine	0	0
Vanillylamine	0	0

* 360 mμ activation; 490 mμ fluorescence.

later than normetanephrine; the amount usually present in urine is so small that further washing of the column before applying a new sample is not necessary unless the presence of large amounts of either is suspected.

Somewhat higher values for the conjugated M and NM content of urine were found than had been reported by some other workers, so the possibility was investigated that falsely high values might have been obtained because of the occurrence of interfering substances. To do this the free and conjugated amines were separated from urine in the usual manner, the eluates from the Amberlite CG-50 column which contained both M and NM were collected together, acidified to pH 4, and amines were again concentrated on Dowex 50. This material was then applied to paper and subjected to chromatography in isopropyl alcohol-ammonia-water (8 : 1 : 1). Several fluorescent spots could be observed when the chromatograms were examined with ultraviolet light, especially a material (R_F 0.17) in the free amine fraction. The chromatograms were then sprayed with a solution of 0.5 per cent potassium meta periodate in borate buffer, pH 5, allowed to stand for 20 minutes, and were dipped in a 0.2 per cent solution of ascorbic acid in 5 N sodium hydroxide. When the chromatograms were again examined under ultraviolet light two strongly fluorescing spots in the positions of M and NM were observed; all the other fluorescent materials had disappeared with the exception of the substance in the free amine fraction, R_F 0.17, the fluorescence of which had markedly weakened but was still visible. This material, however, showed a white fluorescence, distinctively different from the greenish-yellow fluorescence produced from M and NM. The fluorescence spectra of the fluorophore from urinary amines always matched closely those obtained with authentic amines; this indicates that little interference from other substances with different fluorescence characteristics could have occurred.

Finally, the amine content of eight urine samples, selected because the contained unusually high amounts of conjugated amines, was determined by semi-quantitative paper chromatography and also the fluorometric method, and the results were compared. The urine samples were processed in the usual manner, the eluates from the CG-50 column were concentrated on Dowex-50 and the amines were subjected to chromatography in the isopropyl alcohol-ammonia system. Known amounts of authentic amines were processed with the same procedure and were applied to the paper chromatograms as standard reference spots. The intensity of the colors produced from authentic und urinary amines when the chromatograms were sprayed with diazotized p-nitroaniline were compared visually, and an estimate of the amount of amine in the urine was made. Estimates made in this manner by the different methods agreed within 10 per cent, which is the limit of accuracy of the chromatographic method. The above results show that only the M and NM in urine are measured by the fluorometric method.

g) Recovery of amines added to urine. Known amounts (10 μg.) of each amine were added to urine samples which were then processed in the usual manner. The same samples were assayed without the addition of authentic compounds, and the amounts originally present were substracted from the total amounts found in the recovery experiments. The results are listed in Table 2. These data provide an estimate of the reproducibility of the procedure as well as showing the recovery rate.

In Table 2 are listed also data on recovery of added amines when the first sodium acetate washing was omitted. It can be seen that much higher recoveries are obtained in this manner. As noted earlier, however, the sodium acetate wash removes substances which disturb the chromatography of the amines on the CG-50 column; there are usually eluted about 3 ml. earlier when the wash is omitted.

Table 2. *Recovery of authentic amines added to urine*

The metanephrine and normetanephrine contents of urine samples were determined. 10 μg. each of the amines were added to another portion of the same sample, and the procedures were repeated. The lower part of the table lists values obtained when the sodium acetate wash was omitted, and the portion of eluate collected was altered slightly as discussed in the text. The values for recovery of added amines were obtained by subtracting the amounts originally present in the samples from the total amounts measured in the samples to which authentic amines had been added. For the total amines, the urine was hydrolyzed without separating free and conjugated amines.

To test recovery of the conjugated amines, authentic free amines were added to the effluent from the Dowex-50 column which contained the conjugated amines, and the resulting solutions were hydrolyzed in the usual fashion.

Amine added	Number of determinations	Range of recovery	Mean recovery rate and standard deviation
		per cent	per cent
Free metanephrine	9	71— 86	78 $\pm$ 4
Free normetanephrine.	9	67— 78	71 $\pm$ 4
Conjugated metanephrine	9	82— 97	90 $\pm$ 4
Conjugated normetanephrine . .	9	69— 89	78 $\pm$ 6
Total metanephrine	7	91— 98	94 $\pm$ 2.5
Total normetanephrine	7	83— 95	89 $\pm$ 3.5
Free metanephrine	8	91—103	97 $\pm$ 4
Free normetanephrine.	8	88— 93	91 $\pm$ 2
Conjugated metanephrine	8	93— 99	96 $\pm$ 2
Conjugated normetanephrine . .	8	82— 94	87 $\pm$ 4

h) Hydrolysis of conjugated amines. A urine sample was passed through Dowex-50 (H$^+$) to remove free amines and the effluent, which contained conjugated amines, was adjusted to 0.1 N in hydrochloric acid by the addition of concentrated acid. The sample was divided into portions, which were placed on a steam bath and heated different lengths of time, and the M and NM content of each portion was determined in the usual manner. The results of this experiment are presented in Fig. 5. Thirty minutes heating was sufficient to hydrolyze all the conjugated amines; continued heating did not cause a significant loss of amines. In order to be certain that hydrolysis of urea with resultant production of ammonia had not neutralized the acid so that complete hydrolysis did not take place, an additional volume of concentrated acid equal to that originally used was added to one portion after one hour of heating and it was then heated another 30 minutes. No more of either amine was released and there was actually some loss in the more strongly acid solution.

In order to confirm that comparable results are obtained when free and conjugated amines are determined separately or as total amines, 5 different samples of urine were assayed with both procedures. The results, listed in Table 3, show there is good agreement in the total amount

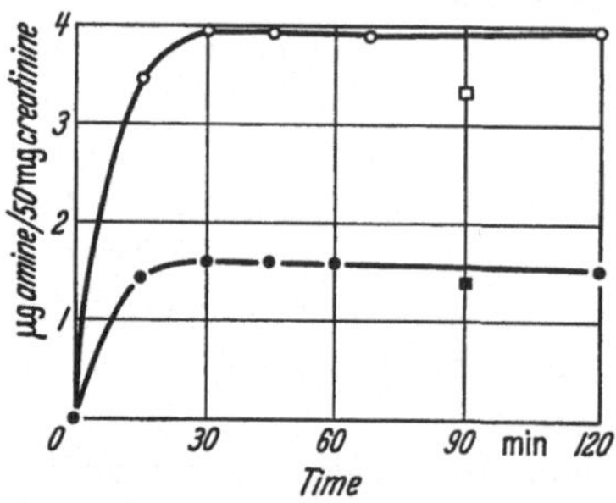

Fig. 5. Release of conjugated amines by acid hydrolysis. The conjugated amine fraction from a urine sample was adjusted to 0.1 N in hydrochloric acid, and heated on a steam bath for the length of times indicated in the figure. Metanephrine and normetanephrine where then estimated in the usual fashion. Closed circles, metanephrine; open circles, normetanephrine. After one portion had been heated 1 hr., an additional amount of acid equal to that already used was added and heating was continued 30 minutes. Metanephrine and normetanephrine were then measured as usual. Closed square, metanephrine; open square, normetanephrine

of each amine as measured by totalling the free and conjugated fractions and by direct determination.

Table 3. *Comparison of amount of total amines measured directly or by adding together free and conjugated fractions*

The samples were divided into two portions. With one portion, free and conjugated amines were measured as described in the text. The other portion was adjusted to 0.1 N in hydrochloric acid, heated 30 minutes on a steam bath, and the total amines were measured. Results are expressed as μg. of amine per 100 mg. creatinine.

Sample	Amine	Free	Con-jugated	Total (sum of free and conjugated)	Total (direct determinations)
1	Metanephrine	2.45	9.00	11.45	11.70
	Normetanephrine	1.03	8.08	9.11	9.50
2	Metanephrine	2.78	10.36	13.14	12.90
	Normetanephrine	1.29	10.10	11.39	10.80
3	Metanephrine	1.44	5.49	6.93	6.95
	Normetanephrine	0.52	6.76	7.28	7.05
4	Metanephrine	1.34	5.53	6.87	6.50
	Normetanephrine	0.81	5.70	6.51	6.43
5	Metanephrine	1.52	6.86	8.38	8.56
	Normetanephrine	0.57	7.43	8.00	8.14

i) Nature of the fluorophores. Adler and Magnusson [11] have shown that guaiacol compounds are rapidly demethylated and then further oxidized by treatment with metaperiodate. Thus, the fluorophores formed from M and NM by periodate oxidation undoubtedly are the same that are produced in the many variations of the trihydroxyindole method for the determination of the catecholamines. The activation and fluorescence spectra of the fluorophore produced from M and epinephrine by the procedure used here were essentially identical, as were those from NM and norepinephrine, except that, on a molar basis, slightly less fluorescence was formed from the O-methyl derivatives. This would be expected, because the formation of methanol from the ethers by the action of periodate is not quantitative [11], and side reactions probably occur. The absorption spectra of the pigments formed by periodate oxidation of the amines were also examined. Larger amounts (0.1 and 0.5 μmoles) of all the amines were subjected to oxidation at pH 5. A red color developed quickly in solutions containing the catechols and more slowly with the O-methyl derivatives. After 40 minutes the absorption spectra of these pigments were measured in both ultraviolet and visible light. Two broad peaks were observed with all the amines. The material produced from epinephrine and M had peaks at 490 mμ., ε, 2.7×10^3 and 1.9×10^3, respectively; and at 305 mμ., ε, 8.6×10^3 and 6.3×10^3. The pigments formed from norepinephrine and NM had peaks at 460 mμ., ε, 1.9×10^3 and 1.6×10^3, respectively, and at 290 mμ., ε, 5.2×10^3 and 4.5×10^3.

VI. Results

The results of analyses of 24 hour urine specimens from 6 healthy adult males are listed in Table 4. The mean value for daily total M excretion (164 μg.) is considerably greater than that reported by SMITH and WEIL-MALHERBE [6] (72 μg.)

or Brunjes *et al.* [7] (110 µg.) but less than that by Yoshinaga *et al.* [4] (220 µg.) and much less than by Kraupp *et al.* [5] (330 µg.). The values for daily total NM excretion (244 µg.) agreed better with the results of the others with the exception

Table 4. *Daily excretion of metanephrine and normetanephrine*
Values are expressed as µg. of amine per 24 hours.

Subject	Wt. (kg)	Metanephrine			Normetanephrine			Total creatinine (mg.)
		Free	Conjugated	Total	Free	Conjugated	Total	
PV	55.9	31.8	141.3	173.1	17.1	225.6	242.7	1760
TK	61.6	32.1	136.9	169.0	13.5	164.0	177.5	2160
KT	52.1	39.6	158.4	198.0	18.1	158.1	176.2	2029
WS	81.5	25.0	97.2	122.2	30.1	232.7	262.8	2465
MA	79.5	25.0	157.7	182.7	28.3	362.2	390.5	2875
SK	58.0	23.2	112.9	136.1	20.0	206.4	226.4	1894
Mean	64.8	29.5	134.1	163.6	21.2	223.2	244.9	2197
±S.D.	±11.5	±5.7	±25.8	±26.3	±5.9	±67.5	±72.0	±376

of Kraupp *et al.* (640 µg.). Despite the fact that the small number of samples analyzed might be responsible for the discrepancy between the results, particular concern was felt because of the lower values for total M excretion reported by Smith and Weil-Malherbe [6] and Brunjes *et al.* [7], who also used a fluorometric method. For this reason the comparison between the amount of amines estimated to be present in the eluate from the CG-50 column by paper chromatography and by the fluorometric method was made, as described earlier, to eliminate the possibility that an interfering substance might have caused falsely high results to be obtained.

Individual values are listed in Table 4 as well as mean values because the use of mean values obscures the evaluation of individual patterns of excretion of the different forms of the amines present. Preliminary studies have indicated that a wide variation occurs in the proportion of free to conjugated forms of both amines, and in the proportion of M to NM in urine from different individuals. Estimations made on many different samples of urine obtained from the same person at the same time of day have shown a considerable degree of constancy in the ratios between the different forms present. Studies on diurnal variations in the excretions of the amines by individuals were also made, and wide variations, as much as from 50 to 150 per cent of the mean value for an individual for the day, were found to occur, as was also observed by Smith and Weil-Malherbe. In confirmation of their observations, no constant pattern of diurnal variation was observed. However, the same individual tended to repeat the same daily pattern of variation. Studies of the significance of individual variations in the proportion of free to conjugated amines and of M to NM excreted, and in the individual patterns of diurnal variation are of obvious interest for future research.

Because 24 hour urine collections cannot always be obtained and are not necessary for some purposes, particularly for the diagnosis of pathological conditions, mean values are listed in Table 5 for the amine content of samples of urine collected from 30 normal adult male subjects during the late morning. These results are expressed on the basis of the creatinine content of the urine samples. For comparison, values are listed in Table 6 for the amine content of urine from patients with surgically confirmed tumors of sympathetic nervous tissue. It is probable that with suitable correlation of values found for urinary free and bound M and NM with

Table 5. *Mean values for excretion of metanephrine and normetanephrine*
Urine samples were collected just before noon from 30 healthy young adult males. Results are expressed as μg. amine per 100 mg. creatinine in sample.

	Mean	Standard deviation	Range
Metanephrine			
Free	1.15	.47	.47—2.22
Conjugated . . .	5.20	1.84	2.29—9.50
Total	6.35	2.22	2.78—11.72
Normetanephrine			
Free	1.26	.58	.48—2.77
Conjugated . . .	8.58	2.36	5.23—13.20
Total	9.84	2.66	5.99—14.69

Table 6. *Excretion of metanephrine and normetanephrine by patients with tumors of sympathetic nervous system tissues*
Values are expressed as μg. per 100 mg. of creatinine in urine sample.

Subject	Type of tumor	Metanephrine			Normetanephrine		
		Free	Conjugated	Total	Free	Conjugated	Total
D. C.	Adrenal pheochromocytoma	1.8	14.0	15.8	103	728	831
H. N.	Adrenal pheochromocytoma	220	2160	2380	124	394	518
C. D.	Adrenal pheochromocytoma	80	584	664	44	246	290
J. T.	Adrenal pheochromocytoma	3.2	15.2	18.4	114	832	946
K. N.	Adrenal pheochromocytoma	4.3	59.2	63.5	505	2530	3035
M. R.	Paraganglioma of celiac plexus	1.4	11.2	12.6	70.5	695	766
D. G.	Pheochromocarcinoma of peri-renal area	2.5	19.5	22.0	52.6	257	310
J. B.	Metastasized adrenal pheochromocytoma	13.2	106.8	120.0	40.8	150	191
M. S.	Neuroblastoma	0.4	2.0	2.4	7.0	54.2	61.2
S. P.	Neuroblastoma	2.6	20.6	23.2	163	1310	1473
L. F.	Neuroblastoma	0.4	17.0	17.4	7.2	47.6	54.8
	Normal (mean and 0) . .	1.15 ±.47	5.20 ±1.84	6.35 ±2.22	1.26 ± .58	8.58 ±2.36	9.84 ±2.66
	(range) . . .	.47 2.22	2.29 —9.50	2.78 —11.72	.48 —2.77	5.23 ±13.20	5.99 —14.69

the clinical information concerning such tumors, these determinations may become of value for the diagnosis of type and location of tumors, in addition to merely indicating their existence [1].

References

[1] EULER, U. S. VON, and I. FLODING: Diagnosis of pheochromocytoma by fluorometric estimation of adrenaline and noradrenaline in urine. Scand. J. Clin. Lab. Invest. 8, 288 (1956).

[2] AXELROD, J., S. SENOH, and B. WITKOP: O-Methylation of catechol amines in vivo. J. Biol. Chem. 233, 697 (1958).

[1] This material has appeared in Journal of Lab. and Clin. Medicine.

[3] Armstrong, M. D., A. McMillan, and K. N. F. Shaw: 3-Methoxy-4-hydroxy-D-mandelic acid, a urinary metabolite of norepinephrine. Biochim. Biophys. Acta 25, 422 (1957).

[4] Yoshinaga, K., C. Itoh, N. Ishida, T. Sato, and Y. Wada: Quantitative determination of metadrenaline and normetadrenaline in normal human urine. Nature 191, 599 (1961).

[5] Kraupp, O., H. Bernheimer, and P. Papistas: Isolierung und Quantitative Bestimmung von 3-O-Methyladrenalin und 3-O-Methylnoradrenalin im Harn. Clin. Chim. Acta 6, 851 (1961).

[6] Smith, E. B., and H. Weil-Malherbe: Metanephrine and normetanephrine in human urine: Method and results. J. Lab. Clin. Med. 60, 212 (1962).

[7] Brunjes, S., D. Wybenga, and J. J. Varner jr.: Fluorometric determination of urinary metanephrine and normetanephrine. Clin. Chem. 10, 1 (1964).

[8] Kakimoto, Y., and M. D. Armstrong: The phenolic amines of human urine. J. Biol. Chem. 237, 208 (1962).

[9] Bertler, A., A. Carlsson, and E. Rosengren: Fluorimetric method for differential estimation of the 3-O-methylated derivatives of adrenalin and Noradrenalin (Metanephrine and Normetanephrine). Clin. Chim. Acta 4, 456 (1959).

[10] Randrup, A.: On the differential fluorometric determination of metadrenaline and normetadrenaline. Clin. Chim. Acta 6, 584 (1961).

[11] Adler, E., and R. Magnusson: Periodate oxidation of phenols. I. Monoethers of pyrocatechol and hydroquinone. Acta Chem. Scand. 13, 505 (1959).

Unité de Biologie Expérimentale, Institut Gustave Roussy, Villejuif (Seine) France

Réflexions sur le Dosage de la Catéchol O Méthyl Transférase et de la Monoamine Oxydase dans les Neuroblastomes

Premiers Résultats Expérimentaux

Par

C. Bohuon, E. H. La Brosse, M. Assicot et A. Amar-Costesec

Les tumeurs nées d'ébauches sympathiques posent aux biochimistes, de délicats problèmes. Mais parmi ceux-ci, il en est un qui a particulièrement retenu notre attention; il s'agit de la présence ou surtout de l'absence d'hypertension chez des enfants porteurs de tumeurs, en apparence identiques tant du point de vue histologique, que du point de vue des métabolites retrouvés dans leurs urines. Il existe un paradoxe frappant entre les quantités énormes d'acide vanillylmandélique retrouvées dans l'urine de certains enfants et l'absence de tout signe pouvant être rattaché à une secrétion accrue d'amines pressives.

Ce problème pourtant intéressant n'a jusqu'ici que très peu retenu l'attention des spécialistes. Et cependant, la grande majorité des enfants porteurs de neuroblastomes n'ont pas d'hypertension. A notre connaissance, seul le travail de Labrosse et Karon [1] a apporté certains arguments en faveur d'une des solutions possibles. Ces auteurs ont pu déceler dans trois tumeurs une activité catéchol O méthyl transférase (COMT). La présence de cet enzyme pourrait, selon eux, suffire à méthyler la fonction phénol en position 3 et donc à inactiver les amines pressives synthétisées au sein de la tumeur.

La question est pourtant à notre avis fort importante, car sa solution permettrait de mieux comprendre l'origine des tumeurs sympathiques de l'enfant. Elle permettrait également de mieux classer les neuroblastomes sur le plan biochimique avec l'espoir de trouver une relation avec l'évolution de la tumeur.

Nous avons donc voulu rechercher et éventuellement doser dans les tumeurs sympathiques, les enzymes, connus depuis déjà un certain nombre d'années et responsables de l'inactivation des catécholamines: la monoamine oxydase (MAO) et la catéchol O méthyl transférase (COMT).

Mais avant d'exposer les résultats de nos investigations dans ce sens, il semble intéressant de réfléchir sur les diverses façons d'aborder ce problème en fonction des diverses hypothèses possibles.

La première hypothèse et la plus simple à imaginer, déjà faite par LABROSSE et coll. [1] et par VON STUDNITZ et coll. [2], est de supposer que les enzymes catabolisants sont présents dans les tumeurs non hypertensives et que la tumeurs secrète dans le sang de la normétanéphrine ou des produits encore plus dégradés par exemple l'acide vanillylmandélique (VMA) ou le méthoxy 3 hydroxy 4 phényl glycol (MHPG).

Nous avons plusieurs arguments en faveur de cette première hypothèse:

a) le travail déjà signalé de LABROSSE et KARON,

b) un autre travail plus récent de LABROSSE et coll. [3]. De la noradrénaline 7-³H est introduite dans les cultures de neuroblastomes. Dans le surnageant après 24 heures on retrouve de la méthoxynoradrénaline 7-³H et surtout du méthoxy 3 hydroxy 4 phényl glcyol 7-³H, ces expériences prouvant que dans ces tumeurs il existe au moins 3 enzymes qui sont mis en évidence en culture de tissu.

— La catéchol O méthyl transférase (COMT)
— La monoamine oxydase (MAO)
— L'aldéhyde réductase (A.R.)

Ces 3 enzymes pourraient par exemple intervenir dans l'ordre suivant:

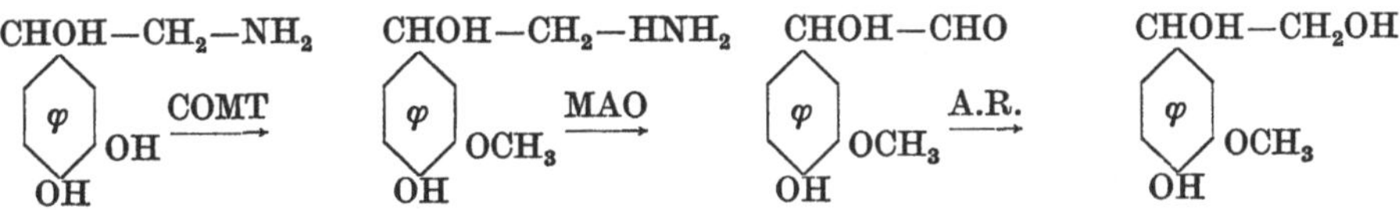

Cependant un travail de SJOERDSMA et coll. [4] a montré dans les phéochromocytomes la présence de MNA[1] et MHPG. Ce résultat obtenu chez des malades où l'hypertension est très nette, prouve que les phénomènes sont complexes et que des schémas trop simples n'expliquent pas tout.

On peut cependant résumer ces deux cas extrêmes dans le tableau suivant. Le type I correspond à plusieurs entités biochimiques (d'après le produit principalement secrété).

Nous verrons d'autre part que nos récentes recherches ne peuvent encore ni confirmer, ni infirmer cette façon d'envisager le problème et nous discuterons, dans la conclusion, le rôle respectif des deux enzymes COMT et MAO dans la transformation in situ de la noradrénaline.

---

[1] MNA = méthoxy nor adrénaline.

	Type	Tissu tumoral	Signes cliniques
I	a	MNA	
	b	MHPG	Pas d' hypertension
	c	VMA	
II		NA Absence d' enzymes MAO et COMT	Hypertension Sueurs etc . . .

Dans certaines tumeurs où il existe surtout une élévation urinaire isolée de HVA, on peut également envisager le schéma suivant:

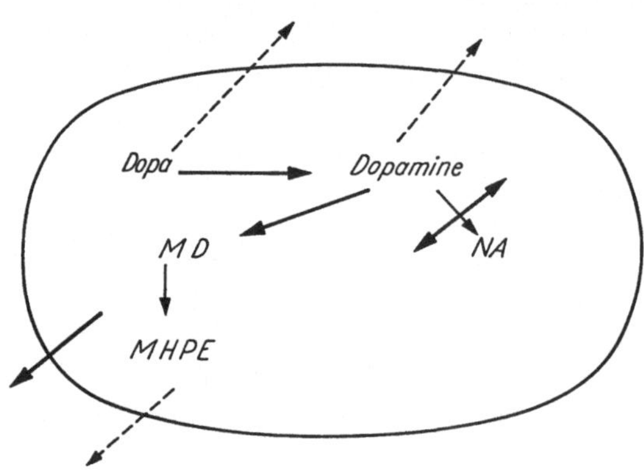

Une autre hypothèse pourrait être la suivante: lorsqu'il n'y a pas d'hypertension, les tumeurs secrètent la noradrénaline sous une forme combinée. Cette combinaison n'aurait aucune action pharmacodynamique classique au niveau des sites récepteurs habituels. La combinaison serait détruite au niveau du foie. C'est peut être ce qu'avaient supposé von Studnitz et coll. [2] en pensant que la réponse vasculaire était modifiée. En fait nous n'avons aucun argument pour penser qu'il en est ainsi.

L'étude des concentrations du plasma en catécholamines déterminées par une méthode fluorométrique, chez les porteurs de tumeurs nerveuses (autres que ceux présentant des signes d'hypertension manifeste) serait en ce sens intéressante.

On peut également penser que toutes les tumeurs ont la même potentialité enzymatique, mais que la libération de la NA à partir des structures de réserve est différente. Ainsi dans les phéochromocytomes. Crout et Sjoerdsma [5] suggèrent que la liaison des catécholamines n'est pas la même que dans les tissus normaux.

Dans ces conditions, des quantités plus ou moins importantes de l'amine pressive peuvent être déversées dans le torrent circulatoire. L'étude au microscope électronique et le dosage de la NA et des activités enzymatiques (MAO et COMT) dans l'ultrafractionnement cellulaire, devraient

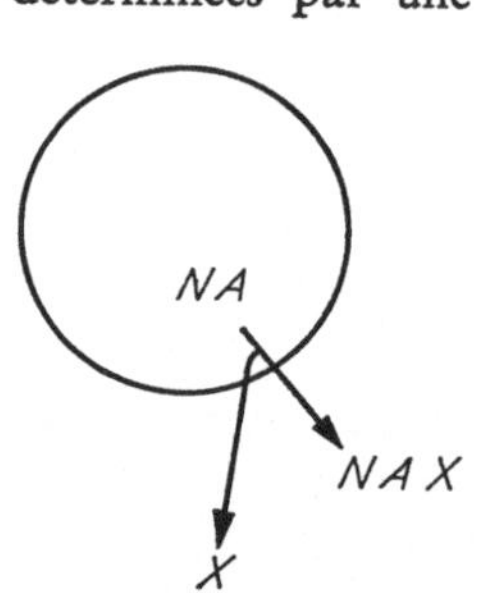

(combinaison inactive sur le plan pharmaco-dynamique) [1]

amener de nouveaux arguments pour ou contre cette hypothèse. Nous comptons étudier certains aspects de cette dernière, en utilisant d'une part les cultures

---

[1] Abréviations: NA = nor adrénaline, MD = méthoxy dopamine, MHPE = méthoxy 3 hydroxy 4 phényl éthanol.

de tissu tumoral et d'autre part les moyens offerts par la microscopie électronique.

Enfin, une autre hypothèse, séduisante mais peu vraisemblable, est la suivante: dans le cas très fréquent où le clinicien ne retrouve aucun signe d'hypertension, la tumeur ne fabriquerait pas de NA (ou du moins des quantités minimes). Il y aurait dans ce cas un «court circuit» dans la fabrication de la NA. La méthoxy noradrénaline et le MHPG seraient secrétés par le tissu tumoral.

On peut imaginer le schéma suivant:

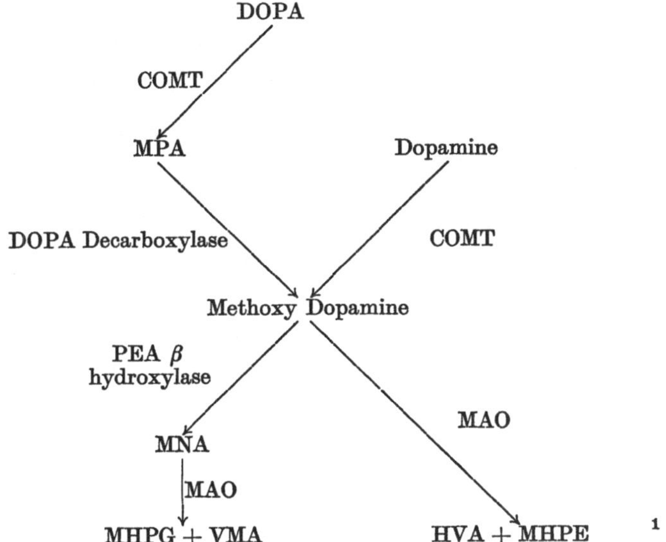

On voit que dans cette hypothèse, il n'est plus nécessaire d'envisager un passage par les amines pressives ou du moins leur excrétion en grande quantité. Mais il faut alors admettre une béta hydroxylation préférentielle de la méthoxy dopamine.

En conclusion, on peut donc penser que la vérité est très complexe et réside peut être dans un mélange de toutes ces hypothèses. Les études biochimiques et cliniques suggèrent fortement qu'il existe plusieurs catégories de neuroblastomes.

Résultats expérimentaux

I. Activité MAO

Un fragment de tumeur (conservé à —20° C) est prélevé, fractionné puis additionné de cinq fois son poids de tampon phosphate 0,05 M à pH 7,7. On homogénéise avec un ultra turax. L'homogénéisat est filtré sur gaze. Sur le filtrat sont dosées la MAO et les protéines.

La MAO a été dosée selon la méthode décrite par WEISSBACH et coll. [6] qui utilise comme substrat la cynuramine. Les incubations sont faites à +37° C dans

---

¹ Abréviations: MPA = méthoxy 3 hydroxy 4 phényl alanine, PEA β hydroxylase = dopamine β hydroxylase, HVA = acide homovanillique.

les cuves thermostatées d'un photomètre Eppendorf. La lecture est réalisée à 366 millimicrons avec enregistrement automatique de la droite: DO = (ft). Un blanc est fait pour chaque dosage afin d'éliminer l'effet de la sédimentation des particules sur la densité optique. L'activité spécifique est exprimée en millimicromôles de cynuramine transformées par minute et par milligramme de protéine.

Les résultats sont exposés dans le tableau 1.

Tableau 1. *Activité MAO dans les tumeurs nerveuses de l'enfant*
N. B. Toutes les tumeurs étudiées correspondent à des malades où aucune hypertension notable n'avait été observée.

BAR	0,69 · mµM/min/mg
COU	2,02 · mµM/min/mg
BRA	1,86 · mµM/min/mg
GRU	0,48 · mµM/min/mg
CAMP	0,56 · mµM/min/mg
AKRI	2,92 · mµM/min/mg

A titre de comparaison, nous donnons aussi les valeurs trouvées pour le foie et le rein chez l'enfant et pour le foie chez le cobaye.

BAR	Foie	2,36 · mµM/min/mg
	Rein	1,94 · mµM/min/mg
COU . . .	Rein	2,62 · mµM/min/mg

Foie de cobaye additionné de 5 volumes d'eau et homogénéisé	3,52 · mµM/min/mg
Mitochondries de foie de cobaye	12,12 · mµM/min/mg

II. Activité COMT

Des fragments de tumeurs (environ 2 g) sont homogénéisés dans 4 ml d'une solution de chlorure de potassium à 1,15 g % à + 4°. L'homogénéisat est centrifugé durant 30 minutes à 14 000 g. L'activité catéchol O méthyl transférase est déterminée sur la fraction surnageante.

Pour le dosage de l'activité catéchol O méthyl transférase nous avons utilisé la méthode décrite par Axelrod [7] légèrement modifiée en ce qui concerne les quantités de substrat et de S adénosyl méthionine [8]. La fraction surnageante de tumeur est mise en incubation à 37° à pH 7,6 en présence de magnésium, de S adénosyl méthionine et de DL adrénaline 7 [3]H. (6 curies/mM, New England Nuclear Corp.).

La réaction enzymatique est bloquée au bout de 90 minutes par du tampon borate à pH 10. La méthoxy 3 adrénaline formée est extraite par un mélange toluène-alcool isoamylique et sa radioactivité est déterminée à l'aide d'un compteur à scintillation liquide.

Des tubes témoins renfermant la fraction surnageante bouillie de tumeur sont mis à incuber dans les mêmes conditions ainsi que la fraction soluble de catéchol O méthyl transférase obtenue à partir du foie de rat (avec une purification d'environ 10 fois).

Les résultats sont exposés dans le tableau 2.

Tableau 2. *Activité COMT dans les tumeurs nerveuses de l'enfant*

	μM de Métanéphrine formée	μM de Métanéphrine formée rapporté au mg de protéines
BAR[1]	$1,87 \cdot 10^{-2}$	$8,31 \cdot 10^{-3}$
COUV	$4,78 \cdot 10^{-3}$	$5,79 \cdot 10^{-3}$
BRA	$3,08 \cdot 10^{-3}$	$4,56 \cdot 10^{-3}$
GRU	$3,86 \cdot 10^{-3}$	$2,71 \cdot 10^{-3}$
CAMP[2]	$6,04 \cdot 10^{-3}$	$4,47 \cdot 10^{-3}$
AKRI	$2,18 \cdot 10^{-3}$	$1,71 \cdot 10^{-3}$
Foie de'un enfant	$5,72 \cdot 10^{-3}$	$3,05 \cdot 10^{-3}$
Foie de rat . . .	$1,13 \cdot 10^{-1}$	$5,02 \cdot 10^{-2}$

[1] présence de MHPG en culture de tissu.
[2] Tumeur d'un type particulier en culture de tissu.

Conclusion

Les études portent encore sur trop peu de tumeurs pour que nous puissions nous faire une opinion définitive. D'autre part nous n'avons pu à l'époque de ce travail étudier les phéochromocytomes, tumeurs qui nous auraient permis d'intéressantes comparaisons sur l'importance de la COMT et de la MAO dans l'inactivation in situ des catécholamines.

Enfin, il est nécessaire d'élargir notre champ d'investigation aux enzymes règlant l'anabolisme des catécholamines (tyrosine hydroxylase et dopamine β hydroxylase).

Quoi qu'il en soit, ces études ont permis de montrer la présence constante de quantités non négligeables de MAO et de COMT dont il faut sans aucun doute tenir compte dans l'explication de l'absence d'hypertension.

Résumé

Dans cette communication les auteurs étudient les diverses hypothèses qui permettent d'expliquer l'absence d'hypertension, très généralement observée chez les porteurs de tumeurs des ébauches sympathiques. Ils étudient les enzymes catabolisants (MAO et COMT) dans les tumeurs mêmes; l'activité enzymatique observée, permet d'expliquer partiellement l'absence d'hypertension. Ils pensent cependant qu'un court-circuitage méthoxy-dopamine, méthoxy-noradrénaline, est une voie de recherche intéressante qui mérite d'être approfondie.

Summary

The authors propose and discuss some hypothesis on the frequent absence of hypertension in children with tumors of neural crest. They give some results on the determination of MAO and COMT in tumors and the ability for the breakdown of norepinephrine. These values can explain partly the clinical observations. The by pass methoxydopamine-normetanephrine is an other interesting explanation, but only hypothetic.

Bibliographie

[1] La Brosse, E. H., et M. Karon: Nature 196, 1222 (1962).
[2] Studnitz, W. von, H. Kaser et A. Sjoerdsma: New Engl. J. Med. 269, 232 (1963).
[3] La Brosse, E. H., J. Belheradek, G. Barski, C. Bohuon et O. Schweisguth: Nature 203, 195 (1964).

[4] Sjoerdsma, A., L. C. Leeper, L. L. Terry et S. Udenfriend: J. Clin. Invest. 38, 31 (1959).
[5] Crout, J. R., et A. Sjoerdsma: J. Clin. Invest. 43, 94 (1964).
[6] Weissbach, H., T. E. Smith, J. W. Daly, B. Witkop et S. Udenfriend: J. Biol. Chem. 235, 1160 (1960).
[7] Axelrod, J., R. W. Albers et C. D. Clemente: J. Neurochem. 5, 68 (1959).
[8] Assicot, M.: Travaux non publiés.

Problèmes posés à l'Organicien par la Synthèse des Métabolites non Azotés des Catécholamines

Par

Jean Gardent et Joseph Likforman

Les principaux problèmes relatifs à la synthèse des catécholamines semblent à l'heure actuelle résolus de façon satisfaisante. La synthèse des acides hydroxy-4 méthoxy-3 mandélique (V.M.A.) et dihydroxy-3,4 mandélique a été décrite en 1958 par Shaw, McMillan et Armstrong qui ont en même temps fait une bonne revue de mise au point sur la synthèse des acides homovanillique et homoprotocatéchique [12]. En 1961 sont décrits l'hydroxy-4 méthoxy-3 phényl éthanol et le dihydroxy-3,4 phényl éthanol [7], en 1963 le dihydroxy-3,4 phényl éthylène glycol et l'hydroxy-4 méthoxy-3 phényl éthylène glycol (M.H.P.G.) [3]. Dans un domaine connexe une synthèse aisément reproductible de l'avide vanilloyl formique a été décrite en 1955 [6].

Un problème reste cependant posé. Les aldéhydes hydroxy-4 méthoxy-3 mandélique (Ia) et dihydroxy-3,4 mandélique (Ib) sont réputées être des intermédiaires dans le catabolisme des catécholamines [1, 8] mais leur synthèse n'a pas encore été réalisée.

$$\text{RO}-\bigcirc\!\!-\!\text{CHOH}-\text{CHO}$$
$$\text{HO}-$$

I a R = CH₃
I b R = H

C'est le problème de la synthèse de l'aldéhyde hydroxy-4 méthoxy-3 mandélique que nous avons abordé sans d'ailleurs y avoir apporté de solution.

Disons d'ailleurs qu'au regard de l'organicien il y a bien peu de chance pour que cet aldéhyde ait une existence durable, aucun aldéhyde mandélique n'ayant à notre connaissance été décrit jusqu'à ce jour. Lorsqu'on tente de préparer les produits de cette espèce trois phénomènes sont en effet suceptibles de se produire.

— Isomérisation en cétol correspondant [5]
— Décomposition en aldéhyde de type aldéhyde benzoïque [9]
— Polymérisation [10, 13].

Le résultat le plus favorable parait avoir été obtenu par Weygand, Betmann Ziemann et Kieger [13] qui décomposent le diéthyl mercaptal de l'aldéhyde benzoïque par le chlorure mercurique en présence de carbonate de cadmium. Ils obtiennent intermédiairement une huile présentant en I.R. une bande carbonyle et une bande hydroxyle mais qui ne tarde pas à se transformer en un polymère solide, dont l'analyse répond à celle de l'aldéhyde mandélique, donnant sous l'action de la dinitro-2,4 phényl hydrazine dans l'acide acétique la bis dinitro-2,4 phénylhydrazone du phényl-glyoxal mais ne présentant plus en I.R. la bande carbonyle.

Nos premiers essais en vue de synthétiser l'aldéhyde hydroxy-4 méthoxy-3 man-délique ont utilisé l'hydroxy-4 méthoxy-3 phényl glyoxal (V) comme matière pre-mière.

Le projet de synthèse consiste à protéger la fonction aldéhyde en l'engageant dans une combinaison, à hydrogéner la fonction cétone en fonction alcool puis à régénérer la fonction aldéhyde. Ceci suppose que le groupe protecteur utilisé soit résistant à l'hydrogénation et suffisamment labile pour permettre une régénération aisée de la fonction aldéhyde à l'état libre.

La synthèse de l'hydroxy-4 méthoxy-3 phényl glyoxal a été décrite selon deux méthodes différentes.

La première (1953) [2] utilise la condensation du chloral avec le gaïacol en milieu alcalin pour donner l'hydroxy-4 méthoxy-3 phényl trichloro méthyl carbinol (III). Par hydrolyse aqueuse de ce dernier on obtient à la suite d'une oxydo-réduc-tion intramoléculaire l'hydroxy-4 méthoxy-3 phényl glyoxal (V).

$$
\begin{array}{ccc}
\mathrm{CH_3O-\!\!\bigcirc\!\!-} & \xrightarrow{\mathrm{Cl_3C-CHO}} & \mathrm{CH_3O-\!\!\bigcirc\!\!-}^{\mathrm{CHOH-Cl_3}} \\
\mathrm{HO-} & & \mathrm{HO-} \\
\mathrm{II} & & \mathrm{III} \\[4pt]
\Big\downarrow{\mathrm{CH_3COOH}\atop\mathrm{BF_3}} & & \Big\downarrow{\mathrm{H_2O}} \\[6pt]
\mathrm{CH_3O-\!\!\bigcirc\!\!-}^{\mathrm{CO-CH_3}} & \xrightarrow{\mathrm{Se\,O_2}} & \mathrm{CH_3O-\!\!\bigcirc\!\!-}^{\mathrm{CO-CHO}} \\
\mathrm{HO-} & & \mathrm{HO-} \\
\mathrm{IV} & & \mathrm{V}
\end{array}
$$

La seconde (1957) [10] utilise l'oxydation sélénieuse de l'hydroxy-4 méthoxy-3 acétophénone (IV). Celle-ci est elle-même classiquement obtenue par transposition de Fries de l'acétyl gaïacol [11]. Mais nous avons trouvé plus expédient de préparer ce corps par action d'acide acétique sur le gaïacol (II) en présence de fluorure de bore.

Nous avons expérimenté les deux techniques de préparation de l'hydroxy-4 mé-thoxy-3 phényl glyoxal qui donnent toutes deux de bons résultats mais la technique d'oxydation sélénieuse est plus avantageuse en ce qu'elle permet d'opérer à des con-centrations plus importantes et donc d'éviter de trop grands volumes.

Bien que ces réactions de préparation du méthoxy-3 hydroxy-4 phényl glyoxal se réalisent dans des conditions de rendement satisfaisantes on peut cependant dire que les difficultés pratiques commencent dès ce stade. De bons rendements ne sont en réalité obtenus que grâce à l'isolement à l'état de combinaison bisulfitique (VI).

A partir de cette dernière il est possible de préparer l'hydrate par décomposition en milieu acide mais cette préparation ne se fait qu'au prix de pertes importantes.

On peut encore après oxydation sélénieuse distiller sous vide l'hydroxy-4 mé-thoxy-3 phényl glyoxal. Mais cette distillation est elle-même délicate à conduire et produit des pertes importantes par résinification.

C'est entre autre pour cette raison que nous avons d'abord tenté d'hydrogéner la combinaison bisulfitique par l'hydroborure de potassium en espérant que la protection apportée serait suffisante. Mais nous avons dû constater qu'il n'en était rien et que la réduction ménagée de la combinaison bisulfitique par l'hydroborure de potassium conduisait non à l'aldéhyde hydroxy-4 méthoxy-3 mandélique désirée mais au cétol isomère correspondant (VII).

$$CH_3O-\underset{HO-}{\overset{CO-CHOH-SO_3K}{\bigcirc}}\quad \xrightarrow{KBH_4}\quad CH_3O-\underset{HO-}{\overset{CO-CH_2OH}{\bigcirc}}$$

VI   VII

$$CH_3O-\underset{HO-}{\overset{CHOH-CH_2OH}{\bigcirc}}$$

VIII

L'obtention de ce cétol n'est d'ailleurs pas sans intérêt puisqu'il permet d'accéder par hydrogénation catalytique en présence de charbon palladié à l'hydroxy-4 méthoxy-3 phényl glycol (VIII) et éventuellement au produit tritié si l'on opère l'hydrogénation par du tritium.

Une autre méthode de protection de la fonction aldéhyde peut-être envisagée consistant à engager celle-ci dans une fonction oxime.

La préparation des aldoximes de produits type phényl glyoxal, plus communément appelées isonitrosocétones, repose habituellement sur la nitrosation des acétophénones. Cette nitrosation est classiquement effectuée soit en milieu acide soit en milieu alcoolique anhydre en présence d'alcoolate alcalin. Dans notre cas une nitrosation en milieu acide était exclue et n'a même pas été tentée la présence d'une fonction phénolique libre devant induire une nitrosation au moins partielle du noyau aromatique. Quant à la nitrosation en présence d'alcoolate de sodium elle se heurtait d'emblée à l'insolubilité dans le milieu du sel de sodium de la fonction phénol. Ceci nous a conduit à essayer la nitrosation après avoir protégé la fonction phénol par un groupement ultérieurement éliminable. Nous avons tenté cette opération en convertissant l'hydroxy-4 méthoxy-3 acétophénone en carbéthoxy-4 méthoxy-3 acétophénone. Nous avons ensuite voulu procéder à la nitrosation de ce corps par le nitrite d'amyle en présence d'alcoolate de sodium mais cette tentative s'est heurtée à un échec complet car, même en opérant dans des conditions très ménagées, l'alcoolate de sodium scinde instantannément la carbéthoxy méthoxy acétophénone en régénérant le phénol libre dont le sel de sodium précipite.

Finalement nous avons réussi à obtenir l'oxime souhaitée (IX) en faisant réagir directement l'hydroxylamine sur l'hydroxy-4 méthoxy-3 phényl glyoxal (V).

$$CH_3O-\underset{HO-}{\overset{CO-CHO}{\bigcirc}}\quad \xrightarrow{NH_2OH}\quad CH_3O-\underset{HO-}{\overset{CO-CH=NOH}{\bigcirc}}$$

Ceci bien que le phényl glyoxal (X) lui-même traité par l'hydroxylamine conduise non à l'oxime mais à un dérivé hétérocyclique du groupe de l'oxadiazine (XI) [4].

Nous avons tenté d'hydrogéner cette oxime par l'hydroborure de potassium mais nous n'avons ainsi obtenu aucun produit isolable à l'état pur.

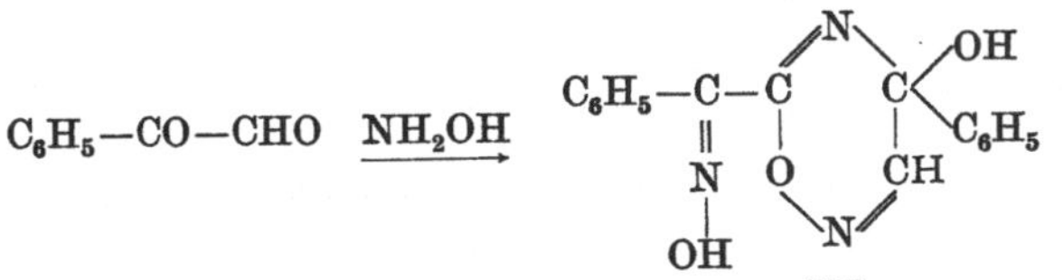

$$C_6H_5-CO-CHO \xrightarrow{NH_2OH} \text{(X → XI)}$$

X   XI

Par contre cette oxime nous a donné une nouvelle voie d'accès simple à l'acide vanillolyl formique. Il suffit pour cela de la traiter par l'anhydride acétique ce qui permet d'obtenir le nitrile acétylé (XII) qui est ensuite facilement hydrolysé en acide vanilloyl formique (XIII) par l'acide chlorhydrique concentré à froid.

Le dernier essai de protection de la fonction aldéhyde que nous ayons tenté a consisté à faire l'acétal méthylique (XIV). L'hydrogénation catalytique (H_2 Pd/C) du produit obtenu permet de préparer l'acétal alcool correspondant (XV). Mais ce

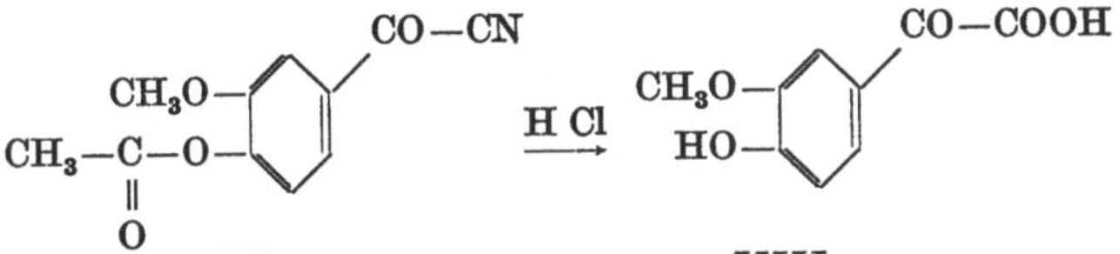

XII   XIII

dernier hydrolysé par l'acide chlorhydrique même dans des conditions extrêmement douces conduit non à l'aldéhyde hydroxy-4 méthoxy-3 mandélique mais à un corps différent. En effet bien que l'analyse élémentaire soit correcte le produit ne présente pas de bande carbonyle en I. R. et doit par conséquent être considéeé comme un polymère du produit désiré.

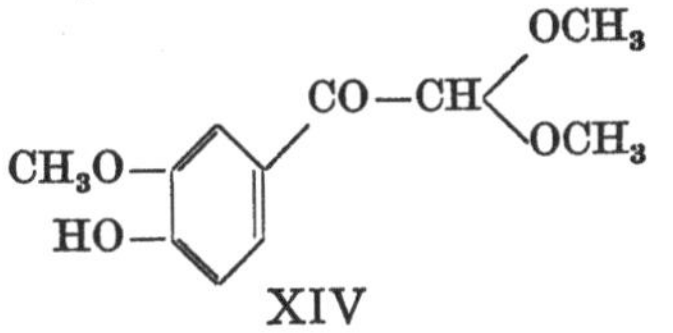

XIV 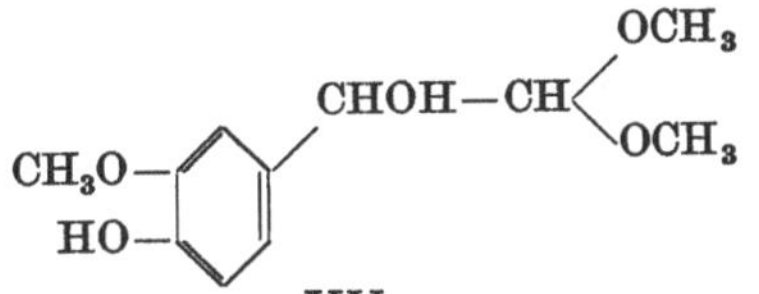   XV

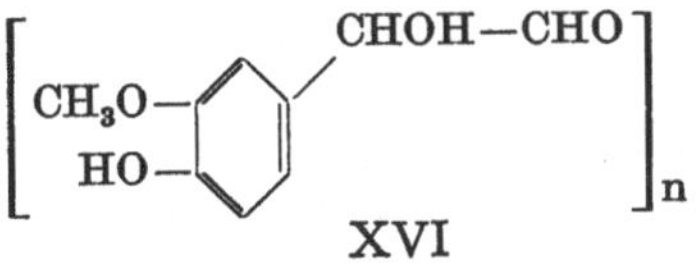

XVI

Les différents essais que nous avons entrepris ne nous ont donc pas permis d'accéder à l'aldéhyde hydroxy méthoxy mandélique. Ils nous ont tout au moins enseigné que les essais ultérieurs n'auront de chance de réussite que si la phase terminale a lieu en milieu neutre ou alcalin.

De plus, chemin faisant, nous avons trouvé deux nouvelles voies d'accès simple à des composés déjà décrits: hydroxy méthoxy phényl glycol et acide vanilloyl formique.

Résumé

Quelques tentatives pour parvenir à l'aldéhyde hydroxy-4 méthoxy-3 mandélique sont rapportées mais on n'a pu isoler jusqu'à présent qu'un polymère de celle-ci ou le cétol isomère. Description de nouvelles méthodes de synthèse de l'hydroxy-4 méthoxy-3 phényl éthylène glycol et de l'acide vanilloyl formique.

Summary

An approach for the synthesis of 4 hydroxy 3 methoxy mandelaldehyde is discussed. At this time, the authors have isolated a polymer of this aldehyde, or the cetol isomer. In this paper, new methods for the preparation of 4 hydroxy 3 methoxy phenyl ethyleneglycol and vaniloyl formic acid are described.

Bibliographie

[1] AXELROD, J., I. J. KOPIN et J. D. MANN: Biochim. Biophys. Acta 36, 576 (1959).
[2] BEKE, D., O. KOVACS, I. FABRICIUS et I. LÄM: Pharmazeut. Zentralhalle 92, 237 (1953).
[3] BENIGNI, J., et A. J. VERBISCAR: J. Med. chem. 6, 607 (1963).
[4] DIELS, O., et E. SASSE: Ber. dtsch. chem. Ges. 40, 4053 (1907).
[5] EVANS, W. L., et C. R. PARKINSON: J. Amer. Chem. Soc. 35, 1770 (1913).
[6] GLENNIE, D. W., H. TECHLENBERG, E. T. REAVILLE et J. L. McCARTHY: J. Amer. Chem. Soc. 77, 2409 (1955).
[7] GOLDSTEIN, M., A. J. FRIEDHOFF, S. POMERANTZ et J. F. CONTRERA: J. Biol. Chem. 236, 1816 (1961).
[8] LEEPER, L. C., H. WEISSBACH et S. UDENFRIEND: Arch. Biochem. Biophys. 77, 417 (1958).
[9] MARSHALL, J. R., et J. WALKER: J. Chem. Soc. 467 (1952).
[10] MOFFET, R. B., B. D. TIFANY, B. D. ASPERGREEN et R. V. HEINZELMAN: J. Amer. Chem. Soc. 79, 1687 (1957).
[11] MOTTERN, H. O.: J. Amer. Chem. Soc. 56, 2107 (1934).
[12] SHAW, K. N. F., A. McMILLAN et M. D. ARMSTRONG: J. org. Chem. 23, 27 (1958).
[13] WEYGAND, F., H. J. BESTMANN, H. ZIEMANN et E. KLIEGER: Ber. dtsch. chem. Ges. 91, 1043 (1958).

Central Laboratory, Dikemark Hospital, Asker, Norway

Cystathioninuria and Vanil-lactic-acid-uria

By

LEIV R. GJESSING

With 2 Figures

The functional neural tumors encompass the neuroblastomas, the ganglioneuromas and the phaeochromocytomas producing dihydroxy-phenyl-alanine and derivatives of this amino acid. These three types of tumors develop from the primitive sympathetic neuroblasts which normally differentiate into nerveganglion cells or into chromaffin cells. The neuroblastoma develops from the primitive neuroblast whereas the ganglioneuroma from the mature ganglion cells and the phaeochromocytoma from the chromaffin cells. The neuroblastoma tissue is frequently mixed with ganglioneuroma tissue. All these three tumors occur in the adrenals and in the sympathetic ganglions in the thorax and in the abdomen as well as on the collum.

We have been interested in estimating the amount of some amino-acids and as many dihydroxy-phenyl-alanine metabolites as possible in the urine and the tumors from patients with sympathetic tissue tumors in order to try to obtain a more differentiated biochemical picture of these tumors. During this study we found 2 new dihydroxy-phenyl-alanine metabolites, namely vanillactic acid [1] and recently also vanilpyruvic acid [2] as well as 2 amino-acids, namely cystathionine

Table 1. *Case material*

Case	Sex	Age	Tumor	Location	Size	Comments
To	♂	68 years	Phaeochromocytoma	Left adrenal	Plum	Successfully operated
Sk	♂	51 years	Phaeochromocytoma	Both adrenals	Tangerine	Died during a crisis
Skia	♂	1 year	Phaeochromocytoma	Both adrenals	2 grams	Died from heart failure
Sa	♀	21 years	Phaeochromocytoma	Fetus?		Recovered after removal of fetus
Ve	♂	4 years	Ganglioneuroma	Abdomen		Alive with his benign tumor
J.R.	♀	18 years	Ganglioneuroma	Abdomen	1200 grams	Successfully operated
Ja	♂	4 years	Neuroblastoma? with glandular metastases	Thorax		Insignificant effect of radiation
He	♂	1 year	Ganglio-neuroblastoma with glandular and orbital metastases	Thorax	Tangerine	Died shortly after operation
Ry	♂	6 weeks	Neuroblastoma?	Neck	Plum	Recovered after removal of the tumor
En	♂	6 months	Neuroblastoma with glandular metastases	Thorax	Orange	Operated but had a recidivation which was sensitive to radiation
St	♀	3 months	Neuroblastoma with glandular metastases and ascites	Abdomen	As big as the fist of a man	Died shortly after operation
Re	♂	9 months	Neuroblastoma with glandular and bone metastases	Right adrenal	Tangerine	Successfully operated
Ei	♂	1 week	Neuroblastoma with extensive metastases	Both adrenals	Filling the whole abdomen	Died 6 days after admission to hospital
Bj	♂	4 years	Neuroblastoma with extensive metastases	Right adrenal	As big as the childs head	Died 2 months after admission to hospital
Ri	♂	3 years	Neuroblastoma with orbital metastases	Left adrenal	Grapefruit	Died shortly after operation
Pe	♀	3 years	Neuroblastoma with extensive metastases	Both adrenals and all organs in the abdom.	All organs in the adbomen involved	Died 8 weeks after the first symptoms

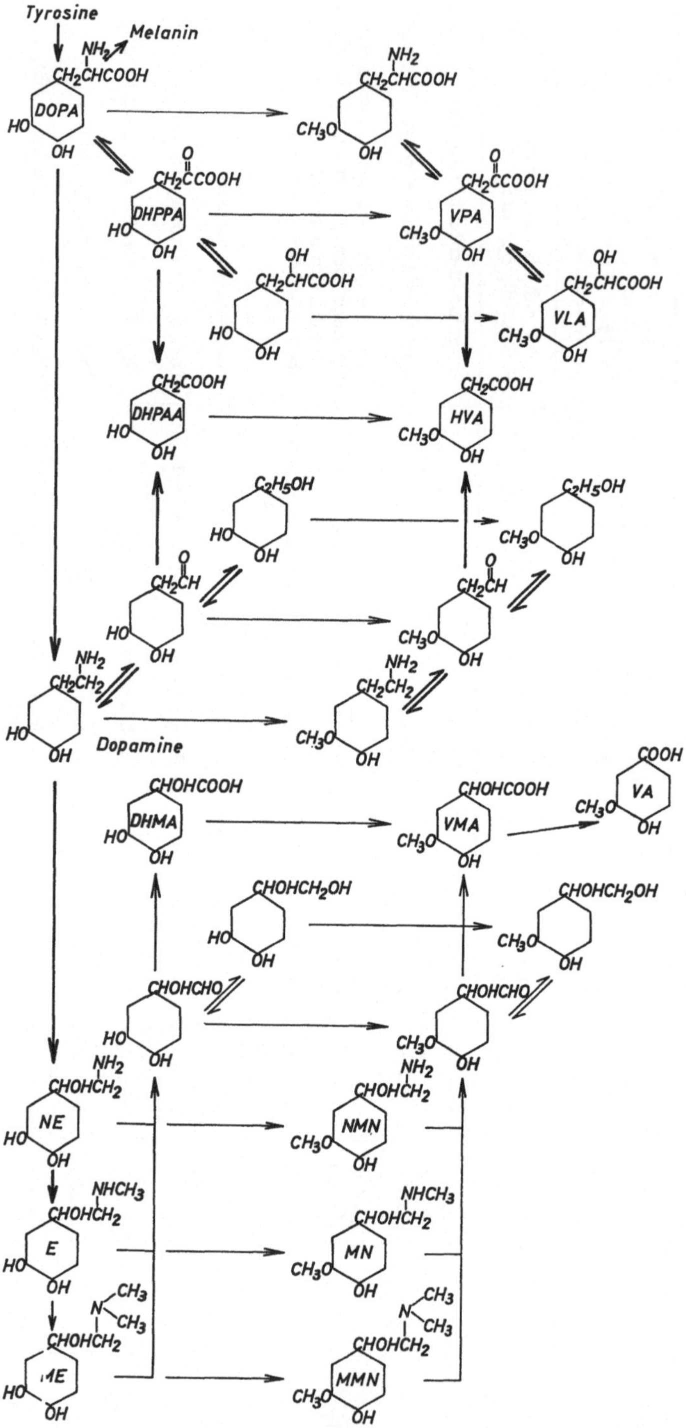

Fig. 1. This scheme summarizes the 12 dihydroxy- and the 14 3-methoxy-4-hydroxy-phenyl metabolites of DOPA

Table 2. *The urinary metabolites from 16 patients with functional neural tumours*

	Phaeochromocytoma				Ganglio-neuroma		Neuroblastoma with glandular metastases					Neuroblastoma with extensive metastases					Normal value[1]	
																	Adults	Children
Case Age Location	$\frac{To}{68}$y A	Sk 51 y AB	$\frac{Skia}{1}$y AB	Sa ?	Ve 4 y Abd	$\frac{J.R.}{18}$y Abd	$\frac{Ja}{4}$y Th	He 1 y Th	Ry 6 w Cer	En 6 m Th	$\frac{St}{3}$m Abd	Re 9 m A	Ei 1 w AB	Bj 4 y A	Ri 3 y A	Pe 3 y A		3 m—12 y
MPA							23	12			20	12	8		30	50		
VLA			3			0.2	14			1	24	2	15	10	7.5	104		
MDA		0.4	0.4	0.3	2	0.3	1	6	0.5		20	5	20	3	9	6	0.06	
MHPE					1.5			13			25	10	50	8				
HVA	3	10	8	6	125	20	375	90	41	70	120	155	480	1000	140	156	2—4	16—3
NMN	0.6	4.4	2.6	7	0.6	3	2	54	7	4	240	10	75	15	0.8		0.1	0.6 —0.05
MN	0.6	3.4	3.3	0.2	0.1				0.2	0.2	0.5		2	2	0.1		0.08	0.16—0.02
MHPG		16.5	25	24	7.5		24	99	60		78	240	750	120	0.8		0.05	
VMA	4	220	30	33	42	5	400	360	68	100	480	297	700	1300	17	15	1.5—3	12—2
VA		40	4	8	4	1	10	15	22	8	24	4	125	80	12		0.2	
CT	—	—	—	—	—	—	5	25	35	140	140	200	800	350	300	250	—	—
β-AIBA	—	—	—	—	—	—	—	—	(+)	+++	+++	++	+++	+	+	+	—	—

All values are expressed in μg. per mg. creatinine. A = adrenal, AB = adrenal bilateral, Abd = abdominal, Th = thoracal, Cer = cervical, y = years, m = months, w = weeks. MPA, VLA, HVA, VMA, CT and β-AIBA are measured as free metabolites, the other 6 metabolites represent the sum of free and conjugated. The 5 new cases are underlined.

[1] Normal values from investigations of 10 men und 13 children.

[3] and β-aminoisobutyric acid [1], in the urine from the cases with malignant neuroblastoma. These 4 compounds seems to be of value for the evaluation of the degree of malignancy of the neuroblastoma as well as for the differential diagnosis between ganglioneuroma — phaeochromocytoma on one hand, and neuroblastoma on the other hand.

1. Methods

The 3-methoxy-4-hydroxy-phenylalanine metabolites were determined with 2 dimensional paperchromatography [4] and the amino-acids with highvoltage paperchromatography combined with ascending paperchromatography [3].

2. Material

We have studied the urinary excretion of ten dihydroxy-phenylalanine derivatives in the urine from 10 patients with neuroblastoma, 2 patients with ganglioneuroma and 4 patients with phaeochromocytoma, see Fig. 1 and Table 1, as well as tumor tissue from 6 cases of neuroblastoma [5, 6], one case of ganlioneuroma and one case of phaeochromocytoma [6] on catecholamines and amino-acids (see Table 3). Recently we have also studied the VPA excretion [2].

The abbreviations used are:

DOPA: dihydroxy-phenylalanine
MPA: 3-methoxy-4-hydroxy-phenylalanine or vanil-alanine
VLA: 3-methoxy-4-hydroxy-phenyllactic acid or vanil-lactic-acid
VPA: 3-methoxy-4-hydroxy-phenylpyruvic acid or vanil-pyruvic acid
HVA: 3-methoxy-4-hydroxy-phenylacetic acid
MDA: 3-methoxy-4-hydroxy-phenylethylamine
MHPE: 3-methoxy-4-hydroxy-phenylethanol
NMN: normetanephrine
MN: metanephrine
MMN: N-methyl-metanephrine
VMA: vanillyl-mandelic acid
MHPG: 3-methoxy-4-hydroxy-phenyl-glycol
VA: vanillic acid
NE: norepinephrine
E: epinephrine
DA: dopamine
CT: cystathionine
β-AIBA: β-amino-isobutyric acid
DHPPA: dihydroxy-phenylpyruvic-acid
DHPAA: dihydroxy-phenyl-acetic-acid
DHMA: dihydroxy-mandelic-acid

3. Results

The main results of our studies are summarized into Tables 2 and 3, and Fig. 2. Table 2 shows the lack of CT and β-AIBA in phaeochromocytoma and ganglioneuroma whereas CT is present in all the cases of neuroblastoma.

In the phaeocromocytoma the NE metabolites predominate whereas in the ganglioneuroma the HVA is 3 to 30 times higher than VMA. Some of the cases with neuroblastoma have also an excretion of the amino-acids MPA and β-AIBA in addition to CT. The 2 most malignant cases of neuroblastoma (case Ri and Pe) had

Table 3. *Cystathionine and catecholamines in tumor tissue*

Case	Tumor	CT µg./g.	Dopamine µg./g.	NE µg./g.	E µg./g.
To	Phaeochromocytoma	—	—	8000	4000
J. R.	Ganglioneuroma	—	—	—	—
St	Neuroblastoma	30	—	—	—
Pe	Neuroblastoma	20	40	25	78
Re	Neuroblastoma	53	20	35	10
Bj	Neuroblastoma	80	6	6	10

especially high values of MPA and the HVA was app. 10 times higher than VMA. In addition case Pe had large amounts of VLA. Cases with large amounts of VLA also excreted large amounts of VPA [2, 6].

The tumor tissue from the benign phaeochromocytoma and ganglioneuroma (Table 3) had no CT whereas the 6 neuroblastoma cases contained 20—80 µg. per g. [5, 6]. On the other hand the phaeochromocytoma tumor contained large amounts of NE and E whereas the ganglioneuroma case contained none and the neuroblastoma cases only 6 to 80 µg/g.

In the urine we also found that CT, VMA and HVA decreased together after operation and after radiation, and increased together during a recurrence, see Fig. 2 [1].

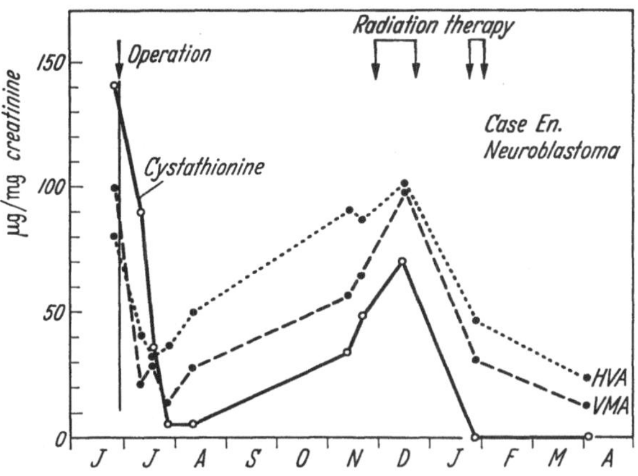

Fig. 2. Case En shows how the urinary excretion of VMA, HVA and cystathionine accompany each other before and after operation as well as during a recurrence and radiation therapy

4. Discussion

From the above mentioned results it seems to be possible to diagnose biochemically the 3 different types of functional neural tumors. The benign phaeochromocytoma excrete predominantly NE or E metabolites without the amino-acids CT, MPA or β-AIBA, the ganglioneuroma excrete more HVA than VMA and none of the above mentioned amino-acids, whereas the malignant neuroblastoma excrete the amino-acids CT, MPA and β-AIBA as well as large amounts of DA and NE+E metabolites. The most malignant cases seems to produce mainly DOPA, MPA, VPA and VLA.

These findings suggest that the least differentiated and most malignant tumors mainly are able to oxidise tyrosine to DOPA which is then methylated to MPA and transaminated to VPA which is partly reduced to VLA and partly decarboxylated oxidatively to HVA. The less malignant neuroblastoma can manage to decarboxylate DOPA to DA and β-hydroxylate DA to NE, whereas the benign phaeochromocytoma have intact enzyme systems producing mainly NE & E metabolites without MPA derivatives and DA.

The increased amount of CT and the low amount of DA, NE and E in the neuroblastoma tumors combined with a large urinary excretion of 3-O-methylated compounds and CT seems to point to a high turnover of catecholamines as well as accumulation and leakage of CT.

In phaeochromocytoma the condition is opposite with a very high concentration of catecholamines, but not CT, in the tumor, but only a slight increase of 3-O-methylated compounds in the urine.

In this case there is an excessive storage of catecholamines in the tissue and a slow liberation of these amines.

The presence of CT in the neuroblastoma seems to be due to insufficient amounts of Vitamin B_6 in the tumor tissue. The increased production of CT through increased 3-O-methylation and the increased demand for Vitamin B_6 for transamination and decarboxylation may explain the elevated level of CT in the tumor tissue. The cystathioninuria is appearantly due to leakage from the tumor cells.

The β-amino-iso-butyric-acid-uria in genetic non excreters has been explained to be a result of extensive cell destruction [8] or to reflect an impairment of liver function [9]. In malignant neuroblastoma the cell destruction may account for this amino-acid-uria.

Summary

1. 4 cases of phaeochromocytoma, 2 cases of ganglioneuroma and 10 cases of neuroblastoma were studied with regard to urinary phenolic acids, alcohols, amines and amino-acids. Tumor tissue from 1 case of phaeochromocytoma, 1 case of ganglioneuroma and 4 cases of neuroblastoma were examined on the content of cystathionine and catecholamines.

2. The determination of cystathionine and β-aminoisobutyric acid, seems to be important for the diagnosis of phaeochromocytoma and ganglioneuroma on one hand and neuroblastoma on the other hand as cystathioninuria and β-aminoiso-butyric-acid-uria are only present in neuroblastoma.

3. The urinary dihydroxy phenylalanine metabolites and the above mentioned amino-acids seems to give valuable information about the degree of malignancy for the neuroblastoma. The metabolites which indicate malignant growth are vanil-alanine or the 3-methoxy-4-hydroxy-phenyl-alanine, vanil-pyruvic acid or 3-methoxy-4-hydroxy-phenyl-pyruvic acid and vanil-lactic acid or 3-methoxy-4-hydroxy-phenyl-lactic acid as well as cystathionine and β-aminoisobutyric acid. The larger the excretion of these 5 metabolites is, the more malignant are the tumors.

4. The content of catecholamines was excessive in phaeochromocytoma tumor tissue, absent in the ganglioneuroma tissue but only present in small amounts in the neuroblastoma tissue. The cystathionine was found only in the neuroblastoma tissue.

Acknowledgement

This work was in part supported by a Research Grant M-5726, National Institute for Mental Health, United States Public Health Service.

The tables and the figures were taken from L. R. Gjessing, Studies of functional neural tumors. VI. Biochemical diagnosis. Scand. J. clin. Lab. Invest. 16, 661 (1964), with the permission of the editor.

References

[1] GJESSING, L. R.: Studies of functional neural tumors. V. Urinary excretion of 3-methoxy-4-hydroxyphenyl-lactic acid. Scand. J. clin. Lab. Invest. 15, 649 (1963).
[2] —, and O. BORUD: Urinary vanilpyruvic acid, neuroblastoma. Lancet 1964 II, 818.
[3] GJESSING, L. R.: Studies of functional neural tumors. II. Cystathioninuria. Scand. J. clin. Lab. Invest. 15, 474 (1963).
[4] — I. Urinary 3-methoxy-4-hydroxy-phenyl-metabolites. Scand. J. clin. Lab. Invest. 15, 463 (1963).
[5] — III. Cystathionine in the tumor tissue. Scand. J. clin. Lab. Invest. 15, 479 (1963).
[6] — VI. Biochemical diagnosis. Scand. J. clin. Lab. Invest. 16, 661 (1964).
[7] RUBINI, J. R., E. P. CRONKITE, V. P. BOND, and T. M. FLIEDNER: Urinary excretion of beta amino *iso* butyric acid (BAIBA) in irradiated human beings. Proc. Soc. exp. Biol. 100/1, 130 (1959).
[8] ARMSTRONG, M. D., K. YATES, Y. KAKIMOTO, K. TANIGUCHI, and T. KAPPE: Excretion of β-aminoisobutyric acid by man. J. Biol. Chem. 238, 1447 (1963).

Swiss Center of Clinical Tumor Investigation (Metabolic Laboratory)

Institute for Clinical Protein Research and Pediatric Clinic, University of Berne
Berne, Switzerland

Mechanism of Action of α-Methyl-Dopa [1]

By
H. KÄSER and H. P. WAGNER

With 3 Figures

Chemically α-methyldopa (AMD) or Aldomet® is the 3,4-dihydroxy-L-phenyl-alanine methylated in the α-position (Fig. 1). SOURKES [28] first showed in 1954 that higher concentrations of this substance in vitro inhibit the decarboxylation of dopa (= 3,4-dihydroxyphenyl-alanine) to dopamine (= 3,4-dihydroxyphenylethylamine), one step in the main pathway of noradrenaline synthesis. This observation, confirmed by DENGLER and REICHEL [8] in animals, and by OATES, SJOERDSMA

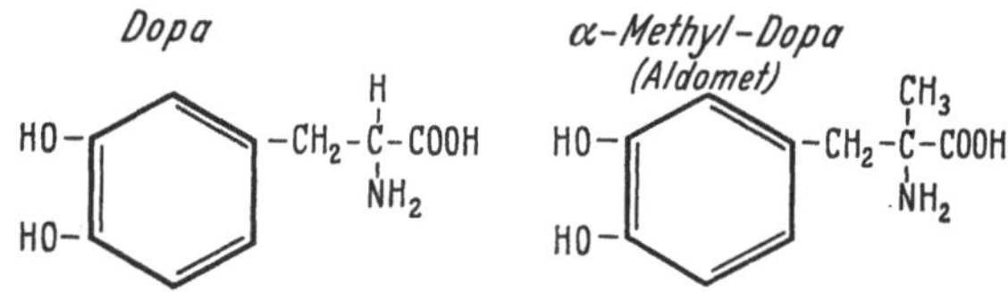

Fig. 1. Structure of Dopa and α-Methyldopa

et al. [16, 25—27] in humans, led to the clinical trial of this substance in hypertensive patients which established its usefullness in the therapy of hypertension [1, 2, 4, 9, 10, 15, 17, 22, 24].

Later on the following facts shed some doubt on the hypothesis that AMD was only a dopa-decarboxylase inhibitor:

After a single oral or intravenous dose of AMD the blood pressure often fell only on the second or third day inspite of the fact that the AMD-concentration in the blood reached low levels 8 hours after administration [22, 27].

---

[1] This work was supported by the Swiss National Foundation for Scientific Research.

AMD analogues with a more pronounced inhibiting action on dopa-decarboxylase in vitro had practically no pressure effect in vivo [4, 10].

After administration of AMD the reduction of the noradrenaline concentration in various organs, such as the central nervous system or the heart, lasts considerably longer than its inhibitory action on dopa-decarboxylase [5, 12, 19]. In addition the blood pressure of hypertensive patients, who did not respond to sympathectomy and who's noradrenaline stores are depleted, is not lowered by AMD [4].

These observations led Hess [12], Gillespie et al. [10] and Pletscher [18] to postulate, that AMD, rather than inhibiting dopa-decarboxylase, had a more reserpine like action and was mainly depleting the tissue stores of the endogenous catecholamines.

Recently however, differences in the mode of action of AMD and reserpine were found: the depletion of noradrenaline by AMD ist not accompanied by any obvious failure of responses to sympathetic nerve stimulation in experimental animals [30] and the effect of a tyramine load after administration of AMD and reserpine is not the same [7]. Since it was also demonstrated in vitro und in vivo [3, 10, 20, 33] that AMD itself may be decarboxylated to a-methyldopamine which in turn is converted to a-methylnoradrenaline, Day and Rand [7] and Scott and Robinson [31] postulated that AMD may inhibit dopadecarboxylase by competition.

In order to obtain additional information on the mode of action of AMD we administered to three healthy adults on a standardized diet 250 mg. Aldomet [1] daily per os and followed the urinary excretion of metanephrine, normetanephrine, vanilmandelic acid and homovanillic acid with chromatographic methods [14, 29]. No effort was made to measure the output of adrenaline and noradrenaline considering the fact that AMD metabolites interfere with the chemical methods used for the determination of these pressor amines [23].

Table 1. *Urinary excretion of catecholamine metabolites in 3 normal adults before and during a-methyl-dopa administration (250 mg./day)*

Catabolite	Individual	Before AMD	With AMD		
			1. day	2. day	3. day
Normetanephrine (μg./day)	A	182	375	256	142
	B	96	336	180	124
	C	112	126	126	108
Metanephrine (μg./day)	A	94	110	86	110
	B	138	124	115	132
	C	125	136	106	136
Vanilmandelic acid (mg./day)	A	2.5	1.8	2.0	2.0
	B	4.1	4.1	3.3	3.6
	C	3.6	1.9	2.7	3.7
Homovanillic acid (mg./day)	A	13.0	7.4	trace	trace
	B	6.4	4.3	3.9	trace
	C	6.6	2.6	trace	trace

Two of the three individuals tested (A and B) had an increased normetanephrine excretion on the first and second day of AMD administration. None had significant changes in metanephrine or vanilmandelic acid output. Homovanillic acid however was greatly reduced to barely measurable amounts on day two or three (Table 1).

---

[1] Aldomet® was kindly supplied by the Research Lab. Div. of Merck, Sharpe & Dohme Inc., New York, represented by Biochimica A.G., Zürich.

In addition two patients with metastatic neuroblastomas showing a slightly elevated blood pressure were tested in the same way in order to obtain data from patients with pathologic catecholamine metabolism (Fig. 2).

During AMD administration the elevated urinary vanilmandelic acid excretion was somewhat reduced in one patient and more so in a second without reaching normal values in either one of them. The homovanillic acid output, which was also

rised before AMD rapidly fell to normal values on the second day. Apart from a slight reduction of the blood pressure no clinical changes were noted.

These results are in accordance with the observations of other authors [6, 13, 21, 22, 31, 32] and seem to indicate, that AMD, at least to some extent, acts as a competitive inhibitor of dopadecarboxylase being metabolized itself to α-methyldopamine and α-methylnoradrenaline, compounds which have been found in tissue and urine [3, 5, 10, 31]. This would explain why the homovanillic acid excretion is reduced, since this phenolic acid cannot be synthesized from α-methyldopamine. In addition the rather large quantities of α-methylnoradrenaline formed from α-methyl-dopamine might displace endogenous noradrenaline from the storage sites a sympathetic nerve endings and lead to a short lasting increase of urinary normetanephrine excretion as seen in our normal adults (Fig. 3). Whether however a competitive dopadecarboxylase inhibition might explain the unchanged

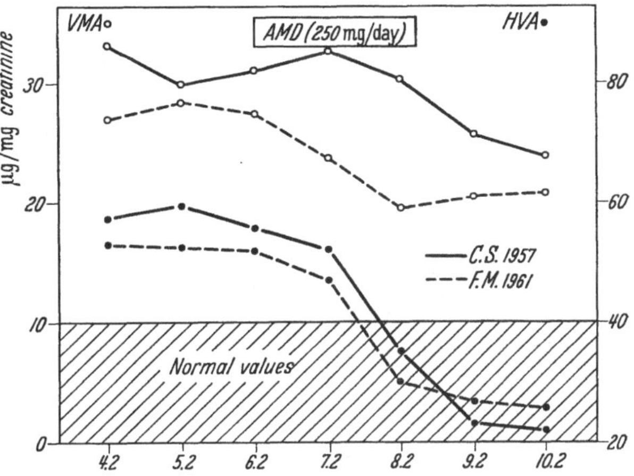

Fig. 2. Urinary excretion of vanilmandelic acid and homovanillic acid in 2 patients with neuroblastoma before and during administration of α-methyldopa

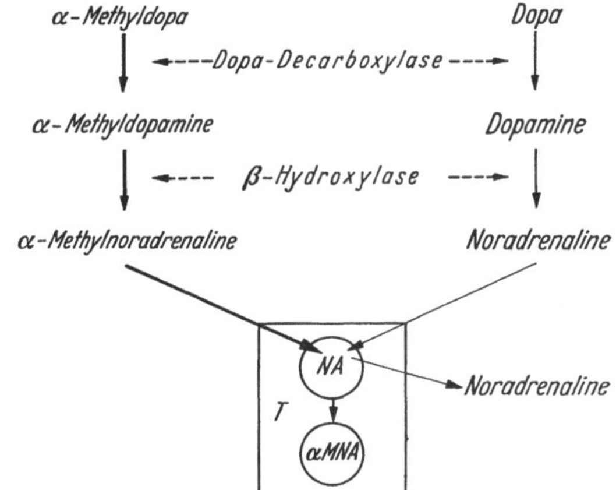

Fig. 3. Postulated effects of α-methyldopa on catecholamine metabolism (T = cell with noradrenaline containing

vanilmandelic acid excretion in the controls and the slight decrease in the patients cannot be said at the present time. If one considers that the controls which were adults and the patients who were children, received the same dose and if one keeps in mind that the catecholamine metabolism of a neuroblastoma differs not only quantitatively but also qualitatively from normal, one might ask, whether any comparison is justifiable under these circumstances. Nevertheless the statement of HESS et al. [12], that the decarboxylation of dopa to dopamine occurs so rapidly that "a 50 per cent reduction of the enzyme activity would result in much less than a 50 per cent reduction of product", might explain why in vivo effects of decarboxylase inhibition become apparent only when the amount of dopa in the tissues is increased. Since patients with neuroblastomas occasionally have an increased urinary dopa excretion [*32*], and may well have abnormally high dopa levels in tissue, the diminution of urinary vanilmandelic acid excretion in these patients might be explained by the mechanisms mentioned above.

Summary

Catecholamine metabolites were determined in the urine of three normal adults and two children with neuroblastomas, all receiving 250 mg. AMD on three consecutive days. Homovanillic acid was markedly decreased in all individuals tested, whereas normetanephrine increased slightly in normals. These results could be explained by competitive inhibition of dopadecarboxylase. However, since vanilmandelic acid seems to be decreased only in patients with neuroblastomas but not in normal controls, other mechanisms might also be responsible for the effects of AMD. In spite of its influence on catecholamine metabolism AMD appears not to be indicated for the therapy of secreting neural tumors.

Zusammenfassung

Es wird über das Verhalten von Katechinaminkataboliten im Urin bei 3 normalen Erwachsenen und 2 Kindern mit Neuroblastom unter kurzfristiger, peroraler Applikation von täglich 250 mg α-Methyldopa berichtet. Dabei konnte bei allen Individuen ein praktisch völliges Verschwinden der Homovanillinsäure beobachtet werden, während das Normetanephrin, das lediglich bei den gesunden Personen bestimmt wurde, einen kurzdauernden, leichten Anstieg zeigte, Resultate, die auf eine kompetitive Hemmung der Dopa-Decarboxylase durch das α-Methyldopa hinweisen könnten. Da die Vanillinmandelsäure jedoch nur bei den Neuroblastompatienten, nicht hingegen den Normalen, etwas abzusinken scheint, müssen aber daneben vermutlich noch andere Mechanismen vorliegen. — Trotz des Einflusses auf die Ausscheidung der Katechinaminkataboliten im Urin scheint die Verabreichung von α-Methyldopa zur Therapie sezernierender, neuraler Tumoren nicht indiziert zu sein.

References

[*1*] BARBOUR, B. H., and M. H. WEIL: Circulation 24, 880 (1961).

[2] BREST, A. N., G. ONESTI, and J. H. MOYER: Circulation 24, 892 (1961).

[*3*] BUHS, R. P., J. L. BECK, O. C. SPETH, J. L. SMITH, N. R. TENNER, P. J. CANNON, and J. H. LARAGH: J. Pharmacol. exp. Ther. 143, 205 (1964).

[*4*] CANNON, P. J., R. T. WHITLOCK, R. C. MORRIS, M. ANGERS, and J. H. LARAGH: J. Amer. med. Ass. 179, 673 (1962).

[*5*] CARLSSON, A., and M. LINDQUIST: Acta physiol. Scand. 54, 87 (1962).

[*6*] CHAPTAL, J., R. JEAN, A. C. DE PAULET et H. BONNET: Press. méd. 71, 833 (1963).

[7] DAY, M. D., and M. J. RAND: J. Pharm. Pharmacol. 15, 221 (1963).

[*8*] DENGLER, H., u. G. REICHEL: Naunyn-Schmiedeberg's Arch. exp. Path. Pharmak. 234, 275 (1958).

[9] GILLESPIE, L.: Ann. N. Y. Acad. Sci. 88, 1011 (1960).

[10] —, J. A. OATES, J. R. CROUT, and A. SJOERDSMA: Circulation 25, 281 (1962).

[11] GREER, M., C. M. WILLIAMS, and L. E. CREVASSE: Clin. Res. 11, 46 (1963).

[12] HESS, S. M., R. H. CONNAMACHER, M. OZAKI, and S. UDENFRIEND: J. Pharmacol. exp. Ther. 134, 129 (1961).

[13] KALLIOMÄKI, J. L., H. A. SAARIMAA, and P. O. SEPPÄLÄ: Curr. ther. Res. 5, 291 (1963).

[14] KÄSER, H., M. EBERHARDT, K. SELLEI u. F. CORNU: Helv. paed. Acta 18, 17 (1963).

[15] LAUWERS, P., M. VERSTRAETE, and J. V. JOOSSENS: Brit. med. J. 1963 I, 295.

[16] OATES, J. A., L. GILLESPIE, S. UDENFRIEND, and A. SJOERDSMA: Science 131, 1890 (1960).

[17] PAYNE, R. W., J. H. CLOSE, T. L. WHITSETT, and J. G. GOGERTY: J. Okla. med. Ass. 54, 430 (1961).

[18] PLETSCHER, A.: Symposium on Clinical Chemistry of Monoamines, pg. 191, London-New York-Amsterdam: Elsevier 1963.

[19] PORTER, C. C., J. A. TOTARO, and C. M. LEIBY: J. Pharmacol. exp. Ther. 134, 139 (1961).

[20] —, and D. C. TITUS: J. Pharmacol. exp. Ther. 139, 77 (1963).

[21] SCHAER, H., u. W. H. ZIEGLER: Klin. Wschr. 40, 959 (1962).

[22] SCHAUB, F., F. NAGER, H. SCHAER, W. ZIEGLER u. P. LICHTLEN: Schweiz. med. Wschr. 92, 620 (1962).

[23] SCHLOSSMANN, K., K. D. BOCK u. G. KRONEBERG: Klin. Wschr. 42, 440 (1964).

[24] SHEPS, S. G., A. SCHIRGER, P. J. OSMUNDSON, and J. F. FAIRBAIRN: J. Amer. med. Ass. 184, 616 (1963).

[25] SJOERDSMA, A.: Ann. N. Y. Acad. Sci. 88, 933 (1960).

[26] —, J. A. OATES, P. ZALTZMAN, and S. UDENFRIEND: New Engl. J. Med. 263, 585 (1960).

[27] — Circulat. Res. 9, 734 (1961).

[28] SOURKES, T. L.: Arch. Biochem. 51, 444 (1954).

[29] SPENGLER, G., H. KÄSER u. G. RIVA: Schweiz. med. Wschr. 93, 1684 (1963).

[30] STONE, C. A., C. A. ROSS, H. C. WENGER, C. T. LUDDEN, J. A. BLESSING, J. A. TOTARO, and C. C. PORTER: J. Pharmacol. exp. Ther. 136, 80 (1962).

[31] STOTT, A. W., and R. ROBINSON: J. Pharm. Pharmacol. 15, 773 (1963).

[32] STUDNITZ, W. v., H. KÄSER, and A. SJOERDSMA: New Engl. J. Med. 269, 232 (1963).

[33] WEISSBACH, H., W. LOVENBERG, and S. UDENFRIEND: Biochem. Biophys. Res. Comm. 3, 225 (1960).

Biochemical Features of Malignant Tumours of Sympathetic Tissues

By

RONALD ROBINSON

With 1 Figure

The neuroblastoma is a primitive and highly malignant tumour. Paradoxically, the phaeochromocytoma, a closely related tumour, is generally benign. Malignant variants occasionally occur, however, and in 1956 MACMILLAN analysed a malignant phaeochromocytoma.

The outstanding finding was that the tumour contained a high concentration of dopamine. MACMILLAN was so impressed with the abundance of dopamine in the tumour that she suggested that the secretion of a very large amount of dopamine may be characteristic of malignant phaeochromocytomas.

If this were true, one should be able to distinguish between a malignant and a benign phaeochromocytoma by a chemical test, namely by measuring the dopamine content of the tumour. If a very high concentration of dopamine were found, this would suggest that the tumour was malignant.

It also seemed possible that one might be able to detect a malignant phaeochromocytoma even before its removal by finding an increased amount of dopamine or its metabolites in the patient's urine.

At this time (1956) we knew very little about the metabolism of catecholamines. The fundamental work of SHAW *et al.* (1957) and other workers has shown that dopa and dopamine are rapidly and effectively metabolized and that homovanillic acid is a major end product of their metabolism. Homovanillic acid is excreted in the urine and it seems reasonable to assume that the rate of its excretion reflects the rate at which dopa and dopamine are being secreted in the body. It should thus be possible to detect dopa and dopamine secreting tumors by demonstrating an increased excretion of homovanillic acid in the urine.

In 1960, three reports were published of greatly increased urinary excretions of homovanillic acid by patients with neuroblastomas (ROBINSON and SMITH; STUDNITZ; GREENBERG and GARDNER). These results suggested that the neuroblastoma was an active secretor of dopa and dopamine.

The striking similarity between the secretory properties of the neuroblastoma and the malignant phaeochromocytoma described by MACMILLAN (1956) was immediately apparent and it seemed possible that the secretion of excessive amounts of dopa and dopamine may be characteristic of malignant tumours of all sympathetic tissues i. e. both sympathetic neural tissue and chromaffin tissue.

To investigate this hypothesis we have examined urine specimens from 65 patients with tumours of sympathetic tissues.

1. Methods and Material

Urine specimens were examined from 16 patients with neuroblastomas, 48 patients with phaeochromocytomas, and from one patient with a carotid body tumour.

Most of the specimens were kindly provided by many colleagues all over Great Britain.

Usually twenty-four hour specimens were collected and an acidified aliquot sent to the laboratory.

Phenolic acids were extracted into ethyl acetate and separated by two dimensional paper chromatography (ROBINSON and SMITH 1959). The chromatograms were inspected visually after they had been sprayed with diazotised p-nitroaniline.

All but a few of the earlier specimens were also examined for phenolic amines. After the urine had been hydrolysed by boiling under reflux at pH 1 for 30 minutes, the specimen was neutralised and the amines were isolated on Dowex-50. They were then eluted with alcoholic ammonia and the solvent was distilled off under reduced pressure. The amines were then separated by two dimensional paper chromatography (ROBINSON and SMITH 1962).

The chromatograms were inspected visually after they had been sprayed with diazotised p-nitroaniline.

Results

Of the 16 patients with neuroblastomas, 15 excreted increased amounts of VMA and homovanillic acid. We also found 4-hydroxy-3-methoxy phenyllactic acid (VLA) in the urine of 7 of these patients.

All these 15 patients excreted excessive amounts of normetanephrine and usually increased amounts of 3-methoxytyramine.

The excessive amounts of homovanillic acid, 3-methoxytyramine, and VLA which these patients were excreting showed that their tumours were discharging large quantities of dopa and dopamine.

Several of the contributors to this symposium have studied larger series of patients than I, and have convincingly shown that neuroblastomas discharge large amounts of dopa and dopamine.

One of our patients excreted only normal amounts of catecholamine metabolites. It is evident that this patient's tumour was non-secretory at least at the time the urine specimen was collected. The patient nevertheless died from the effects of the tumour.

Of our 48 patients with phaeochromocytomas, one had a malignant tumour. We have published a full report of this patient elsewhere (ROBINSON *et al.* 1964). The following account is a summary of our findings.

Just before the tumour was removed, the patient was excreting very large amounts of the phenolic acids VMA, homovanillic acid, VLA, and vanillic acid. The amine chromatogram showed that he excreted large amounts of normetanephrine, metanephrine, N-methyl metanephrine, and 3-methoxytyramine.

The large amounts of VLA, homovanillic acid and 3-methoxytyramine showed that the tumour secreted dopa and dopamine. Despite these findings, there was no histological evidence that the tumour was malignant and my clinical colleagues believed it was benign.

After the tumour was removed, the patient's excretion of catecholamines and catecholamine metabolites became normal and remained so for 11 months. At the end of this time, we noticed that the patient's excretion of VMA was beginning to rise.

A month later, his excretion of both VMA and normetanephrine were significantly higher than normal.

It was evident that the original tumour had been malignant had metastasized, and that the secondary growths were functionally active. Clinical observations agreed with the biochemical findings.

We were thus presented with the opportunity of separately studying the secretory activities of the primary tumour and of the secondary growths an opportunity which we eagerly grasped.

The patient rapidly deteriorated. Just before he died, he was excreting enormous amounts of VMA, VLA, and homovanillic acid. The phenolic acid chromatogram at this time closely resembled that we obtained just before the primary tumour was removed.

The study of this patient was particularly rewarding as it showed that both the primary tumour and the secondary growths secreted large amounts of dopa and dopamine.

There was one outstanding difference between the secretory activities of the primary growth and the secondaries: the primary tumour secreted abundant amounts of the N-methylated amines adrenaline and N-methyl adrenaline, but the secondary growths seemed to be unable to N-methylate noradrenaline.

After the primary tumour had been removed, we detected only normal amounts of metanephrine in the urine and we could not detect N-methyl metanephrine at all.

Why were the secondary growths unable to convert noradrenaline to adrenaline? It is interesting to speculate on the significance of this finding.

In man, only mature chromaffin cells seem to be able to N-methylate noradrenaline. The early foetal adrenal medulla, for example, does not have this property and it contains only noradrenaline. At birth about 30% of the total catecholamines in the gland is adrenaline. The proportion of adrenaline in the gland gradually rises until by the age of about 3 years, 70—80% of the catecholamines in the gland is adrenaline.

It seems reasonable to assume that the tumour cell may mature in the same way that the normal chromaffin cell matures. A rapidly growing tumour or a tumour of recent origin should then consist mainly of immature, noradrenaline-secreting cells, while a longstanding tumour would be expected to contain many mature cells which secrete adrenaline.

The findings on our patient are in accord with this hypothesis. His clinical history showed that his primary tumour had probably been present for about 12 years and it contained many cells which secreted adrenaline; on the other hand, the secondary growths, which were by comparison juvenile, did not secrete adrenaline.

The remaining 47 phaeochromocytoma patients had tumours which, as far as we have been able to ascertain, were benign. None of these patients excreted large amounts of homovanillic acid though one or two patients with tumours which secreted almost exclusively noradrenaline, excreted slightly increased amounts of homovanillic acid. One patient also excreted a small amount of VLA.

The outstanding finding of these studies was that the secretory properties of this malignant phaeochromocytoma closely resembled those of the neuroblastomas in that both tumours secreted large amounts of dopa and dopamine.

Recently, we examined a urine specimen from a man who 18 months earlier had had a malignant carotid body tumour removed. The tumour had metastasized and the patient had secondary growths in the femur and probably in the liver. The urine contained increased amounts of VMA, homovanillic, normetanephrine and 3-methoxytyramine.

This showed that these malignant growths too secreted dopa and dopamine. It is unfortunate that a urine specimen from this patient was not examined before the primary tumour was removed.

The secretory properties of these malignant secondaries contrast with those of the carotid body tumour examined by GLENNER et al. (1962). These workers detected no dopamine in the tumour and there was no evidence that the tumour was malignant.

Discussion

The evidence adduced above is in accord with the suggestion that malignant tumours of sympathetic tissues secrete large amounts of dopa and dopamine. There is, however, one type of tumour, namely the ganglioneuroma, about which the evidence is conflicting.

BELL (1962) found that a patient with a ganglioneuroma — a benign tumour — excreted increased amounts of homovanillic acid indicating that the tumour secreted dopamine; in contrast, the 6 ganglioneuroma patients studied by KÄSER *et al.* (1964) all excreted normal amount of homovanillic acid.

It seems unwise to attempt to generalise about the secretory properties of the ganglioneuroma. Tumours of sympathetic ganglia are frequently of mixea type and may contain cells of all degrees of maturity. Large, apparently mature and benign tumours may contain small neuroblastomatous elements. These more primitive cell-groups may secrete large amounts of dopa and dopamine and yet remain undetected by histological techniques.

It seems generally to be true that malignant tumours of sympathetic tissues discharge significantly larger amounts of dopa and dopamine than benign tumours. The question arises wheter these malignant tumours have specific secretory properties which distinguish them from benign tumours.

A consideration of the kinetics of the reactions involved in the biosynthesis of noradrenaline show that it is not necessary to attribute to the malignant tumours any aberrant secretory properties.

The conversion of dopamine to noradrenaline is a comparatively slow reaction (see Fig. 1). Nevertheless, although it is slow, it does not prevent a high proportion of the dopamine formed in normal sympathetic tissue from being converted to noradrenaline.

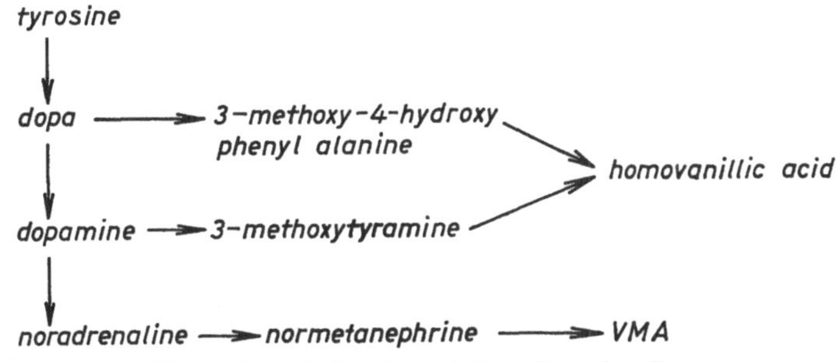

Fig. 1. Biosynthesis and metabolism of noradrenaline

However, in extremely active neoplastic tissue, in which very large amounts of dopa and dopamine are being formed, this reaction may become a bottleneck that prevents most of the dopamine being formed from being converted into noradrenaline.

Dopamine will accumulate in the tumour and inhibit the activity of dopa decarboxylase. Eventually, the tumour will become saturated with both dopa and dopamine which it will then begin to discharge. These compounds and their metabolites will then be excreted in the urin.

Summary

A study has been made of the urinary excretion of catecholamine metabolites of 65 patients with tumours of sympathetic neural and chromaffin tissue.

The patients with malignant tumours excreted significantly larger amounts of dopa and dopamine metabolites than those patients with benign tumours.

It is unnecessary to postulate that these malignant tumours possess aberrant secretory properties: their excessive secretion of dopa and dopamine is attributable to the β-hydroxylation of dopamine being a slow reaction.

References

BELL, M.: In Symposium on Clinical Chemistry of Monoamines. Amsterdam: Elsevier 1963, p. 82.

GLENNER, G. C., J. R. CROUT, and W. C. ROBERTS: Arch. Path. 73, 263 (1962).

GREENBERG, R. E., and L. I. GARDNER: J. Clin. Invest. 39, 1729 (1960).

KÄSER, H., M. BETTEX, and W. VON STUDNITZ: Arch. Dis. Childh. 39, 168 (1964).

MACMILLAN, M.: Lancet 1956 II, 284.

SHAW, K. N. F., A. MCMILLAN, and M. D. ARMSTRONG: J. Biol. Chem. 226, 255 (1957).

STUDNITZ, W. VON: Scand. J. Clin. Lab. Invest., Suppl. 48 (1960).

ROBINSON, R., P. SMITH, and J. RATCLIFFE: J. Clin. Path. 12, 541 (1959).

— — Nature 186, 240 (1960).

— — Clin. Chim. Acta 7, 29 (1962).

—, S. R. F. WHITTAKER, and P. SMITH: Brit. Med. J. 1964 I, 1422.

Observations on the Biochemical Diagnosis of Neuroblastoma

By

MAURICE BELL

With 4 Figures

During the past 4 years intensive work by many investigators has shown the variety of catecholamines and their metabolites which may be excreted in the urine of patients suffering from neuroblastoma. This work was reviewed up to 1962 at the Manchester Symposium (BELL [1]) and major additions since have included VON STUDNITZ, KÄSER and SJOERDSMA [2] and GJESSING [3]. As with many clinical biochemical problems there are two main lines of approach, one leading towards easier diagnosis using simple techniques aimed at major metabolites, the other towards a better understanding of the underlying biochemical abnormality by estimating as many of the metabolites as possible. It is hoped that both may lead to earlier and more specific diagnosis with increased possibility of cure but the emphasis of ones work tends to be towards one extreme or the other. In my case it is towards simpler diagnosis and first I will outline two simple methods which were suggested for this purpose in 1961 (BELL and STEWARD [4]) then illustrate a possible grouping of cases according to their excretion pattern.

1. Screening Tests in the Diagnosis of Neuroblastoma

There is no doubt that no single catecholamine or metabolite is of 100% diagnostic significance even in the chemically positive cases of neuroblastoma reported in the literature. Early emphasis was towards noradrenaline and/or 4-hydroxy-3-methoxy mandelic acid (HMMA, VMA) but in 1961 BELL and STEWARD [4] reported the first 2 cases where the determination of dopamine was of greater diagnostic significance than any of the other metabolites routinely estimated at that time. It was also noted that the "Gitlow" Screening Test [5] for excess HMMA excretion could give grossly positive values in some cases of neuroblastoma. It is the value of these two methods which I will now assess.

2. Materials and Methods

The urine specimens were received from patients referred to the Manchester University Children's Tumour Registry, the Christie Hospital and Holt Radium Institute, the Royal Manchester Children's Hospital and the Alder Hey Children's Hospital, Liverpool and were all investigated between October 1960 and the present time. Random specimens of urine, collected in acid, were received in most cases and all results have been expressed in terms of creatinine. Altogether 26 chemically positive cases of neuroblastoma have been investigated together with 66 cases of other types of tumour in children of similar age group and these include 11 cases where there was some doubt in the histological diagnosis between Ewing's tumour and neuroblastoma.

3. The "Gitlow" Screening Test [5] for Excess HMMA Excretion

A volume of urine equivalent to 0.5 mg. of creatinine is acidified with 5 N hydrochloric acid. It is then extracted 3 times with ethyl acetate, the extracts are combined and evaporated to dryness. The residue is dissolved in water and potassium carbonate added followed by diazotised p-nitraniline. The coloured complex is extracted into amyl alcohol containing 1% ethanolamine and read against water at 450 and 550 mμ. using a suitable absorptiometer. The result is expressed as a ratio of Optical Density 450/Optical Density 550.

Table 1 shows the comparison between the result obtained by this method for 66 cases of non neuroblastoma tumours and the 18 cases of neuroblastoma giving positive values for this test. Two points require emphasis.

Table 1. *Gitlow screening test*
Comparison of results obtained for 66 cases of non neuroblastoma tumour with 18 cases of neuroblastoma giving the positive test.

	Optical density at 550 mμ.			Ratio $\dfrac{OD\,450}{OD\,550}$	
	No.	Mean	Range	Mean	Range
Non neuroblastoma	66	0.19	0.095—0.35	1.58	1.2—2.15
Neuroblastoma	18	2.2	0.54 —8.0	0.63	0.49—1.0

1. Normal adults on dietary restrictions give ratios greater than 1.5 and below 1.3 is suggestive of phaeochromocytoma. It can be seen here that the range for these random specimens from children 1.2—2.15. The 18 cases of neuroblastoma gave ratios between 0.49 and 1.0.

2. The ratio decreases because of the increase of the HMMA complex with its peak reading at 550 mμ. and if the actual readings at this wavelength are considered the striking difference between the two sets of results becomes more evident. The highest reading for a negative result ist 0.35 the lowest for a case of neuroblastoma is 0.54 but 12 of the 18 cases giving positive results had readings of over 1.0 with the highest equivalent to 8.0 for the standard test. (See also Table 3.)

There is no doubt that this test, requiring only a few minutes work, will be positive in the majority of cases of neuroblastoma and a ratio of 1 or less coupled with an optical density of 0.5 or more clearly identifies these case.

4. Screening Method for Total Catecholamines [6, 7]

A suitable aliquot of urine (10 ml. or equivalent to 10 mg. of creatinine) is adjusted to a pH of 6.5 and allowed to drip slowly through a column of Amberlite IRC 50 cation exchange resin buffered at the same pH. After washing the column with water the adsorbed amines are eluted into acid. Ethylenediamine hydrochloride and ammonia are added to a suitable aliquot (usually 5 ml.) and the mixture is incubated at 50° C for 20 minutes. Standards containing 2.5 and 4.0 μg. of dopamine are prepared at the same time and after incubation the mixture is saturated with sodium chloride and the fluorescent product is extracted into isobutanol. Visual comparison can be made using a suitable source of UV light emitting at 365 mμ. If a fluorimeter is available activation at 436 mμ. and reading at 530 mμ. gives less background interference.

Table 2. *Catecholamine screening test*
Comparison of results obtained for 64 cases of non neuroblastoma tumour with 8 cases of neuroblastoma giving the negative "Gitlow" Test.

	No.	Catecholamines (as Dopamine) μg. per mg. of creatinine	
		Mean	Range
Non neuroblastoma	64	0.78	0.23—1.5
Neuroblastoma (Negative „Gitlow")	8		10—230

Table 2 shows the results obtained for 64 cases of non neuroblastoma tumour compared with the 8 cases of neuroblastoma giving the negative "Gitlow" test. The negative cases gave values of less than 1.5 μg. per mg. of creatinine although two not listed here had values of 2.0 and 2.2 μg. respectively. In contrast the lowest neuroblastoma level under these circumstances was 10 μg. per mg. of creatinine. For 6 of these 8 cases the fluorescence was visible in the aqueous phase during the course of the incubation and no extraction or fluorimetric comparison was necessary. whenever the upper standard is exceeded a blank must be carried out on a further 5 ml. aliquot of the same eluate by adding ammonia only before incubating. If the original fluorescence were due to catecholamine it will have disappeared completely from the blank, if due to extraneous material fluorescing in alkaline solution it will still be present and the test is negative.

Table 3 illustrates the combined results for the 26 cases of neuroblastoma investigated. The cases have been separated into groups which I will describe later. In the first group total catecholamines and HMMA are increased in all cases although

Table 3. *Combined results for the total Catecholamine and "Gitlow" screening tests for the 26 cases of neuroblastoma investigated*

Case No.	Total Catecholamines (as Dopamine)[1]	HMMA GITLOW [5]			PISANO [8][1]
		Optical 540 mμ.	Density 550 mμ.	Ratio OD 450/OD 550	
Group 1.					
1.	4.5	1.45	2.75	0.52	140
4.	19	2.08	3.08	0.68	310
5.	9.2	3.92	8.0	0.49	580
6.	10.8	0.95	1.75	0.54	100
7.	26	0.72	0.7	1.0	73
10.	64	0.87	1.135	0.77	76
11.	12.5	0.4	0.3	1.33	11.8
12.	18	1.07	2.15	0.5	110
13.	12.4	1.0	1.7	0.59	74
16.	15	2.3	4.4	0.53	200
21.	35	1.92	3.72	0.54	140
24.	6.2	0.73	1.26	0.58	60
25.	3.4	1.11	2.1	0.53	110
26.	12	0.55	0.91	0.6	48
Group 2.					
2.	120	0.56	0.28	2.0	5.6
3.	232	0.44	0.21	2.1	15
9.	95	0.54	0.37	1.5	24
17.	83	0.36	0.27	1.3	5.8
19.	115	0.19	0.09	2.1	25.5
20.	40	0.45	0.25	1.8	12
22.	10	0.55	0.4	1.38	7.5
Group 3.					
8.	1.6	0.56	1.05	0.53	63
14.	1.8	0.42	0.54	0.8	18
15.	1.8	0.63	0.87	0.73	26
18.	1.25	0.84	0.95	0.87	33
23.	1.5	0.38	0.56	0.68	32

[1] The results are expressed as μg. per mg. of creatinine for the total catecholamines and HMMA by the Pisano method [8].

[2] In the "Gitlow" method when the reading at 550 mμ. exceeded 1.5 less extract was taken and the readings actually obtained multiplied by the dilution factor so that all figures given refer to the equivalent of 0.5 mg. of creatinine.

the "Gitlow" test is negative in No. 11. This table shows the difficulty of using this test alone for diagnostic purposes and illustrates the variations that can occur in the ratio when appreciable amounts of other acid metabolites are present in addition to HMMA. In group 2 cases it is consistently negative although the quantitative values for HMMA are as high as 25 μg. per mg. of creatinine. By contrast one case in group 3 with a quantitative level of only 18 μg. per mg. of creatinine has a ratio of 0.8.

5. The Variety of Excretion Pattern and possible Grouping of Cases of Neuroblastoma

In view of the widespread interest in the possible chemical classification of cases of neuroblastoma according to their excretion pattern all positive cases were further investigated for quantitative excretion of HMMA by the method of Pisano, Crout and Abraham [8] and metadrenaline + normetadrenaline (MA + NMA) by the method of Pisano [9]. Phenolic acid chromatograms were also obtained using a composite method applied to unhydrolysed urine [1]. These determinations were also carried out on all specimens of urine obtained from patients where there was some doubt regarding the histological diagnosis. To show the diagnostic significance of each metabolite more clearly the ratio of the result obtained to that of the upper limit of normal for each metabolite has been plotted. These have been taken as follows: Total catecholamines (as Dopamine) 1.5 μg., MA + NMA 1.0 μg. and HMMA 10 μg. per mg. of creatinine respectively.

a) Group 1

Results for the 14 cases which comprise this group are shown in Figure 1 and in each case there is an increase of all the metabolites estimated here. The total catecholamines are most significantly increased in 5 cases (Fig. 1a.), MA + NMA in 4 (Fig. 1b.) and HMMA in 5 (Fig. 1c.) but there are wide variations and no consistent pattern. The extremes are well illustrated here and the maximum values for total catecholamines, MA + NMA and HMMA are 64, 107 and 580 μg. per mg. of creatinine respectively. The phenolic acid chromatograms showed marked increases in both homovanillic acid and HMMA.

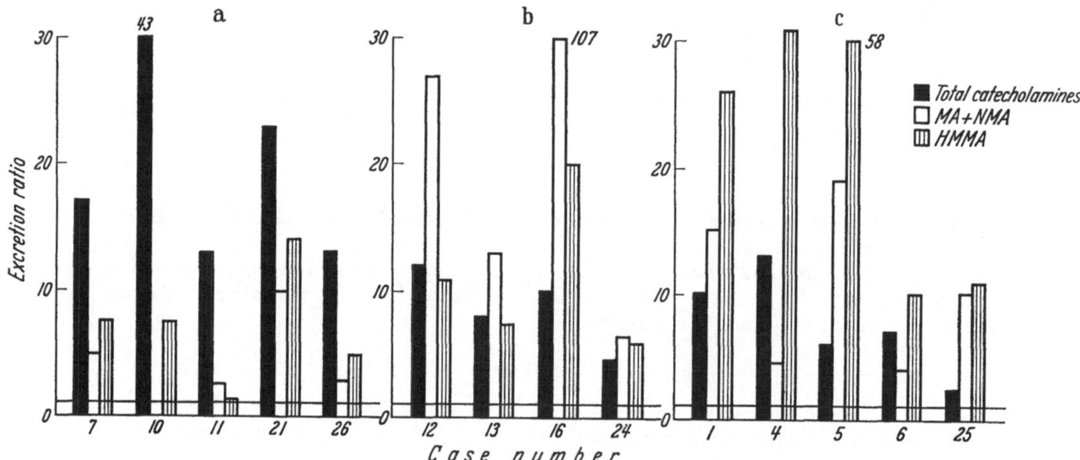

Fig. 1a, 1b and 1c. Patients in Group 1

Fig. 1, 2 and 3. Comparison of the total catecholamine, MA+NMA and HMMA excretion for the 26 cases investigated. The excretion ratios represent the actual values obtained expressed as a multiple of the upper limit of normal (————). These have been taken as follows: total catecholamines (as dopamine), 1.5 μg. per mg. of creatinine; MA+NMA, 1.0 μg. per mg. of creatinine; HMMA, 10 μg. per mg. of creatinine

b) Group 2

As I wrote on an earlier occasion [1] the cases shown in Fig. 2 are perhaps
the most interesting of the series. Here there is an increase in total catecholamines
far in excess of that usually met with in the other groups. The HMMA is either
normal or only slightly increased
and although some MA+NMA
results have been included it is felt
that these are not reliable as there
was marked evidence of appreciable
background interference in the
method not usually met with in
the other groups. Phenolic acid
chromatograms for the first 6 of
these cases show the presence of
large amounts of homovanillic acid
but not the gross amounts of
HMMA usually found in group 1.
They all show the presence of a
large spot adjacent to the HMMA
giving a blue grey colour with
p-nitraniline. At the Manchester

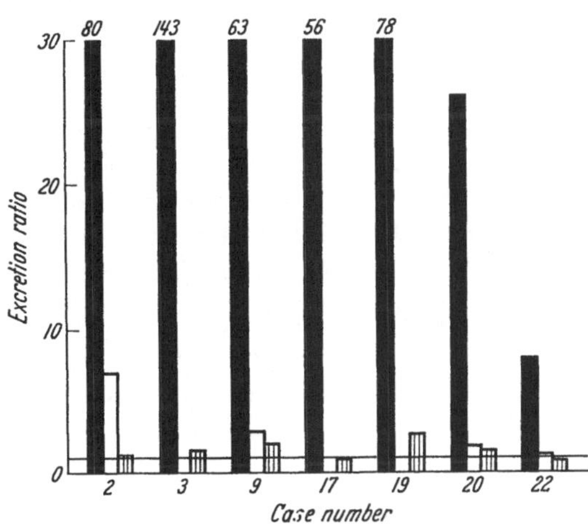

Fig. 2. Patients in Group 2

Symposium in 1962 SMITH and ROBINSON suggested to me that this was 4-hydroxy-
3-methoxy phenyl lactic acid (HMPLA) and it certainly has the characteristics
described for this material by GJESSING [10]. It has not been seen on the chromatograms, obtained under these conditions, for
cases in any of the other groups even though on
occasions the homovanillic acid was present in
much greater concentration.

c) Group 3

This comprises the 5 cases shown in Fig. 3.
characterised by a low catecholamine excretion
and variable increases in MA+NMA and
HMMA. The phenolic acid chromatograms
show little increase of homovanillic acid but
variable increases of HMMA.

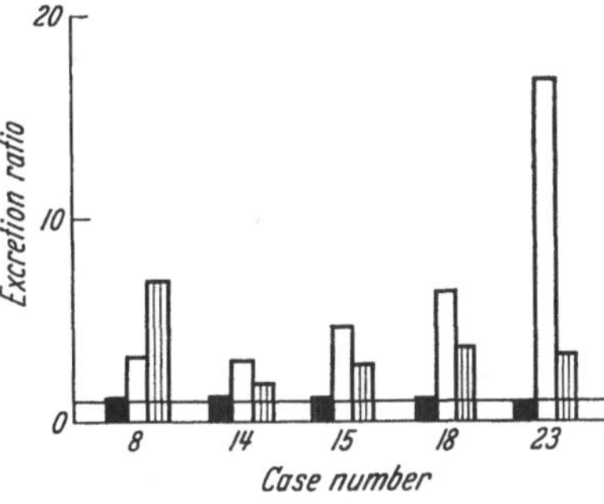

Fig. 3. Patients in Group 3

d) Group 4. ? Ewing's Tumour

At various times cases of neuroblastoma with negative chemistry have been
described and during the course of these investigations 8 cases have been studied
where there was active tumour present but the chemical tests were negative. These
were described as neuroblastoma by some members of the registry panel and by
various names by other members. In none of them was there a firm diagnosis of
neuroblastoma from all histologists and as there were clinical and pathological
grounds for considering them as Ewing's tumours they will not be included here

as chemically negative cases of neuroblastoma. MARSDEN commented on the difficulty of this problem at the Manchester Symposium [11] and he has since added further evidence for the separation of the two types of tumour.

6. Discussion

Diagnosis

Attention has already been drawn to the fact that no single catecholamine or metabolite is 100% reliable in the diagnosis of neuroblastoma. If routine screening is to be undertaken any combination chosen should combine simplicity with a high degree of reliability. This is certainly the case with the two methods recommended here and Table 4 shows the distribution of the positive cases investigated according

Table 4. *Distribution of cases investigated according to screening test result and suggested group*

Group	Total	Screening Test	
		Catechol-amines	Gitlow
1.	14	14	13
2.	7	7	—
3.	5	—	5
	26	21	18
4. ?Ewing's Tumour	11	—	—

to the group and screening test. The catecholamine screening test was positive in 21 cases (80%) and "Gitlow" in 18 (69%).

Turning now to the group 4, Ewing's tumour cases chemically negative values in neuroblastoma have been reported at various times but the number of 8 with active tumour in a series of this size would seem to be rather excessive. VOORHESS *et al.* [12] have reported most negative cases in the literature and of 5 listed in 1963 3 were over 10 years of age. Fig. 4 shows the age distribution of 26 cases of neuroblastoma and 11 cases of ? Ewing's tumour investigated. None of the neuroblastoma children were over the age of 10 years and 19 (73%) were under the age of 5 years. In contrast in the Ewing's tumour group 4 (35%) were over 10 years of age and and only 2 (18%) under 5.

Table 5. *Summary of cases investigated*

Group	Total	Age of onset (Years)		Dead [1]	Living [1]
		Under 2	Over 2		
1.	14	4	10	11 (4.3)	3
2.	7	4	3	7 (5.5)	
3.	5	4	1	1 (2)	4 (23)
4. ?Ewing's Tumour	11		11	8 (22)	3 (45)

[1] Figures in parenthesis show the average time of survival in months after onset of disease.

Table 5 lists the cases investigated according to their group, age on onset of the disease and time of survival. In group 1 the average time of survival was 4.3 months with a range from 1—18 months, group 2 average 5.5 months with a range of

1—13 months and the group 3 survivors average 23 months with a range from 18—30 months. By contrast in the Ewing's tumour group 8 have died but the *average* time of survival was longer than that of any case of neuroblastoma which subsequently died being 22 months and ranged from 8—58 months. Thus the age of onset, prolonged time of survival with the disease and negative biochemistry would appear to separate the Ewing's tumour cases from those of neuroblastoma.

With all the information now at our disposal it would seem to me very doubtful if such a thing as a chemically negative neuroblastoma exists. If tumour tissue is so poorly differentiated how can

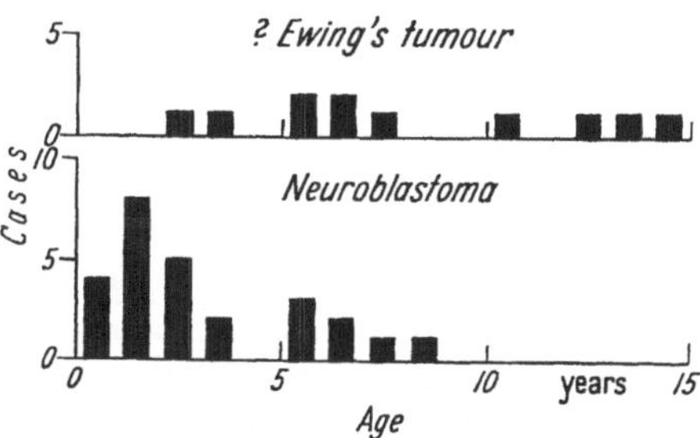

Fig. 4. Age incidence of 26 cases of neuroblastoma an 11 cases of ? Ewing's tumour

anyone say what it might ultimately develope into. In my experience large leukaemoid masses have been queried as neuroblastoma by experienced paediatric pathologists and marrow aspirates from neuroblastoma secondaries described as reticulum cell leukaemia by experience haematologists. Under these circumstances where increased amounts of the dopa/dopamine metabolites on the one hand or noradrenaline metabolites on the other are not detected it is felt that the diagnosis of neuroblastoma is untenable without unanimous agreement on the histological diagnosis by a panel of experience pathologists.

Finally I feel that more cases will be correctly diagnosed at paediatric departments of peripheral hospitals and referred to larger centres if they are made aware of the simplicity and reliability of methods now available. I know of one such hospital where 4 cases of neuroblastoma have been successfully diagnosed in the last 4 years using these methods. On the other hand we have seen no cases from hospitals of similar size and standing where the understanding of the problem is not so great and where they are not prepared, presumably, to carry out these investigations on a screening basis on children with unexplained anaemia, lethargy, bone pain or atypical leukaemic type picture in peripheral blood or bone marrow. One such case had been treated for several weeks for rheumatic fever before being seen by a member of the tumor registry. The urine specimen that followed gave the highest HMMA excretion seen during the course of these investigations.

7. Classification

Returning now to the possible classification of cases of neuroblastoma according to their biochemical excretion pattern it must be possible to correlate the group with some marked biochemical difference, tumour pathology or prognosis. Thus there is little point in subdividing the group 1 cases into those with most significant increases in total catecholamines, MA + NMA or HMMA. These represent minor variations of pattern from tissue which probably contains catechol-O-methyl transferase and monoamine oxidase in addition to those enzymes usually present in sympathetic tissue resulting in the production of a mixture of catecholamines, methylated amines and methylated acid derivatives. This would appear to be the major group reported in the literature.

The position is entirely different with the group 2 cases, or at least with 6 of the 7 of them. Here there would appear to be a deficiency of dopa decarboxylase compared with normal sympathetic tissue. The dopa liberated may be metabolised peripherally either by decarboxylation in the kidney leading to large amounts of dopamine in the urine or by methylation to 4-hydroxy-3-methoxy phenylalanine followed by transamination to the pyruvic and lactic acid derivatives as recently suggested by GJESSING [10]. Some of the pyruvic acid derivative could be decarboxylated to homovanillic which could therefore arise by a pathway independent of dopamine. This possibility could explain the observation that in this group the relative increase of dopamine over normal levels is much greater than the corresponding ratio for homovanillic acid observed in this series or reported in the literature. This would appear to be a somewhat less common variant although it is extremely difficult to assess the incidence in the literature unless dopamine levels are recorded or reference made to 4-hydroxy-3-methoxy phenyl lactic acid. One case reported by VOORHESS and GARDNER [13] undoubtedly belongs to this group as does the one more recently reported by GJESSING [10]. Work on the enzymes present in tumour tissue from cases in this group would be necessary to substantiate this hypothesis and confirm this as a genuine subgroup.

The third group shown here, with one exception, may also be a genuine subgroup. There is the possibility that they may have been less active tumours 2 of which were situated well away from the abdomen, one in the cervical region of the spine the other on the rear wall of the pharynx, but the other 3 were abdominal with local secondaries at least. It is also agreed that the younger the child on diagnosis the better the chances of survival and 4 of these children were under the age of 2 when originally diagnosed (Table 5). In the whole series, however, 12 cases were under this age on original diagnosis, 8 have died with the longest period of survival of 13 months, 4 are still alive with an average survival of 23 months and they are all in this group which may be correlated with survival.

Summary

1. Two methods have been presented which combined give 100% coverage for the diagnosis of neuroblastoma. The simplest, the "Gitlow" test for HMMA, was positive in 70% of the confirmed cases investigated, the other for total catecholamines, positive in 80%.

2. It is suggested that the chemically negative neuroblastoma does not exist.

3. The cases investigated have been classified biochemically as follows.

Group 1. Major increase in metabolites of dopamine and of noradrenaline.

Group 2. Major increase of dopa/dopamine metabolites possibly associated with tissue enzyme deficiency.

Group 3. Increase of metabolites of noradrenaline but not of dopamine possibly correlated with survival.

Acknowledgements

I wish to thank all the persons who have been of assistance to me in the collection of the clinical data and urine specimens for the cases reviewed. In this respect my thanks are due mainly to Dr. J. K. Steward of the Manchester University Children's Tumour Registry, Dr. Dorothy Pearson of the Christie Hospital and Holt Radium Institute, Dr. H. B. Marsden of the Royal Manchester Children's Hospital and Mr. J. T. Ireland of the Alder Hey Children's Hospital, Liverpool.

References

[1] Bell, M.: The Clinical Chemistry of Monoamines. Amsterdam: Elsevier 1963, p. 82.
[2] Studnitz, W. von, H. Käser, and A. Sjoerdsma: New Engl. J. Med. 269, 233 (1963).
[3] Gjessing, L. R.: Scand. J. clin. Lab. Invest. 15, 463 (1963).
[4] Bell, M., and J. K. Steward: Lancet 1961 I, 1061.
[5] Gitlow, S. E., L. Ornstein, M. Mendlowitz, S. Khassis, and E. Kruk: Amer. J. Med. 28, 921 (1960).
[6] Bell, M., R. H. Horrocks, and H. Varley: In H. Varley (Ed.): Practical Clinical Biochemistry. 2nd. ed. London: Heinemann 1958, p. 527.
[7] — M. Sc. Thesis, Victoria University of Manchester, 1960.
[8] Pisano, J. J., J. R. Crout, and D. Abraham: Clin. chim. Acta 7, 285 (1962).
[9] — Clin. chim. Acta 5, 406 (1960).
[10] Gjessing, L. R.: Scand. J. Clin. Lab. Invest. 15, 649 (1963).
[11] Marsden, H. B.: The Clinical Chemistry of Monoamines. Amsterdam: Elsevier 1963, p. 71.
[12] Voorhess, Mary L., L. K. Pickett, and L. I. Gardner: Amer. J. Surg. 106, 33 (1963).
[13] —, and L. I. Gardner: J. Clin. endocrinol. Metab. 22, 126 (1962).

Formation et Renouvellement de l'Acide 3-Méthoxy-4-hydroxymandélique dans les Neuroblastomes

Par

E. H. La Brosse

Avec 4 Figures

Le problème qui m'intéresse est celui des tumeurs de la crête neurale, et de leur métabolisme in vivo. Ces tumeurs, comme on le sait, synthétisent particulièrement la noradrénaline et les malades excrètent les métabolites de celle-ci dans leurs urines. Mais quel est le siège de la formation des produits excrétés? Est-il dans la tumeur elle-même, ou dans un autre tissu?

Les réponses à ces questions ne sont pas faciles parce que le métabolisme de la noradrénaline est très compliqué, comme vous le voyez dans la Fig. 1. Il y a sept métabolites, et plusieurs autres si nous tenons compte des conjugués. Pour cette raison les études que j'ai déjà faites ont montré qu'il est nécessaire d'étudier le renouvellement des métabolites individuellement. J'ai donc commencé par étudier l'acide 3 méthoxy-4-hydroxymandélique (VMA). Ce métabolite a été identifié depuis longtemps par Armstrong [1] et aujourd'hui ce produit est le mieux connu.

Selon toute évidence, d'après les investigations faites, le VMA est excrété libre dans l'urine. Les méthodes pour son isolement sont très bien connues. Nous avons utilisé l'extraction et la chromatographie sur papier pour isoler le VMA, suivant les indications de Armstrong et coll. [2].

Pour étudier la cinétique du VMA j'ai synthétisé du VMA marqué au tritium que nous avons injecté par voie I.V. Nous avons recueilli le VMA dans les urines

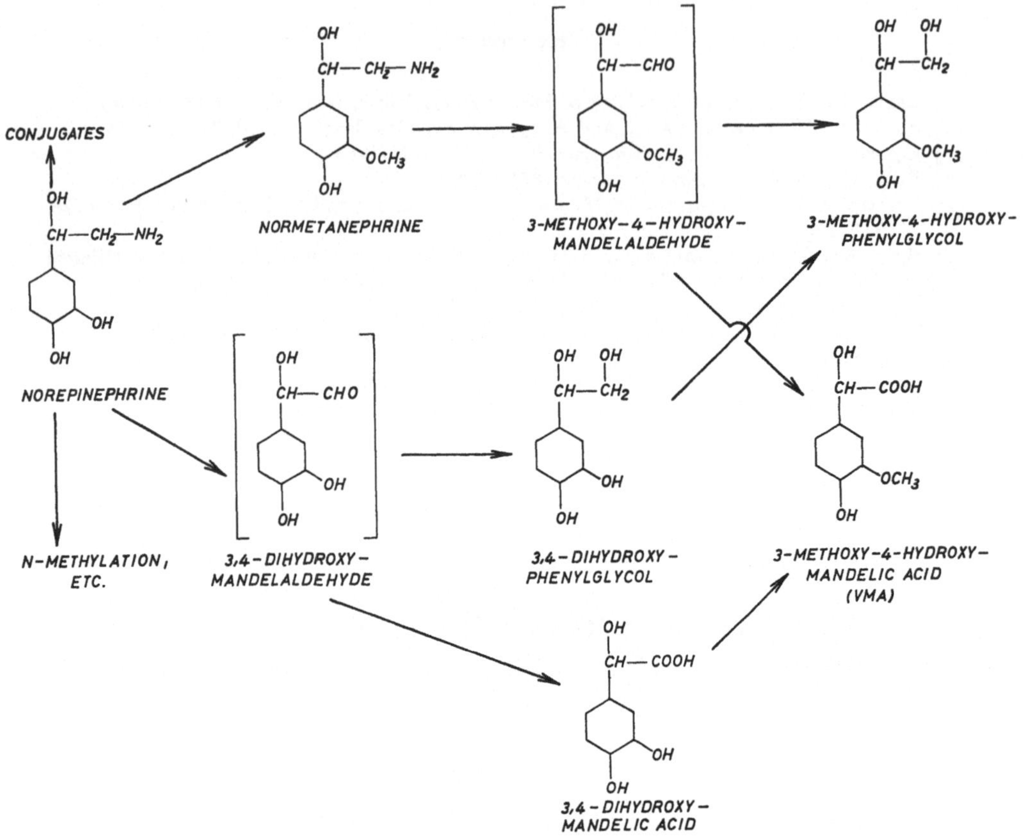

Fig. 1. Voies métabolique pour la norepinéphrine

et dosé l'activité spécifique par [1] des dosages du tritium en scintillation liquide et [2] le VMA par colorimétrie avec la p-nitroaniline diazotée. J'ai étudié 3 sujets normaux et 1 malade aux Etats Unis et nous avons étudié 4 malades avec le VMA marqué, en France.

Les résultats du premier sujet étudié en France ont montré que j'avais donné une dose non traceuse dans tous les cas précédents. J'ai eu la chance de synthétiser grâce à l'assistance de la Section des Molécules Marquées (Centre d'Etudes Nucléaires, Saclay), le VMA avec une très haute activité spécifique. Avec une dose traceuse, j'ai observé un changement dans l'urine qui indique que le ³H-VMA est distribué dans trois compartiments [3].

Dans un deuxième cas, il s'agissait d'un malade avec un neuroblastome thoracique et le VMA urinaire était approximativement 10 fois plus important que la normale. Le changement de l'activité spécifique pour le VMA urinaire est présenté dans la figure 3. Dans ce cas, vous voyez aussi qu'il y a 3 compartiments. Nous avons pensé que peut être le compartiment le plus grand correspondait à la tumeur elle-même et nous avons dosé le VMA dans la tumeur de ce malade. Il y avait seulement

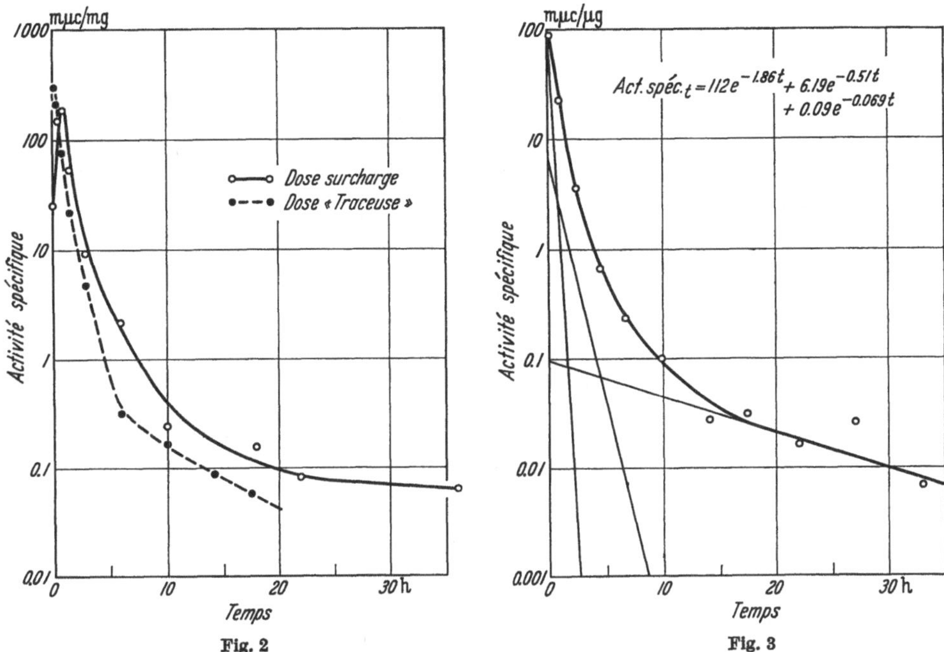

Fig. 2. Changement dans l'activité spécifique du VMA urinaire du malade M.M. après ●—● une dose non traceuse o——o une dose „traceuse", 1/50 de l'autre

Fig. 3. Changement dans l'activité spécifique du VMA urinaire du malade M.L.B. après une dose „traceuse". Ce changement indique trois compartiments

0.07 μg de VMA par gramme de tumeur et non 56 μg/gr de tumeur comme le calcul le faisait prévior. Ces résultats indiquent que la tumeur ne produit pas le VMA directement.

Chez un autre malade nous avons étudié le VMA dans le rein et nous avons trouvé 620 μg/gr. Si nous utilisons la vitesse d'excrétion qui est de 28 mg par jour soit 1170 μg/hr, nous trouvons un renouvellement égal à 0.047/hr.

Pour étudier la clearance rénale nous avons injecté du VMA marqué chez un enfant avec une tumeur de Wilms; lors de l'opération, on a prélevé du sang dans l'artère et dans la veine rénales; le dosage a indiqué que la clearance était de 20%.

Quelle est l'origine du VMA? L'étude en culture de tissu de la tumeur indique que c'est le 3-méthoxy-4-hydroxyphénylglycol qui est secrété (figure 4). Dans ce cas nous avons incubé la 7-³H-noradrénaline avec un neuroblastome en culture de tissu. Après 24 heures le milieu de culture a été prélevé puis on a réalisé un chromato-gramme sur papier. La noradrénaline marquée est changée en 3-méthoxy-4-hydroxy-phénylglycol (MHPG) et non en VMA [4]. Un autre travail que j'ai fait a montré que le MHPG injecté dans le sang est changé très rapidement en VMA.

Voici les conclusions de notre travail.

1° Le VMA est rapidement excrété dans l'urine et son augmentation indique la présence d'un neuroblastome et ces modifications peuvent aider à juger l'effet d'un traitement.

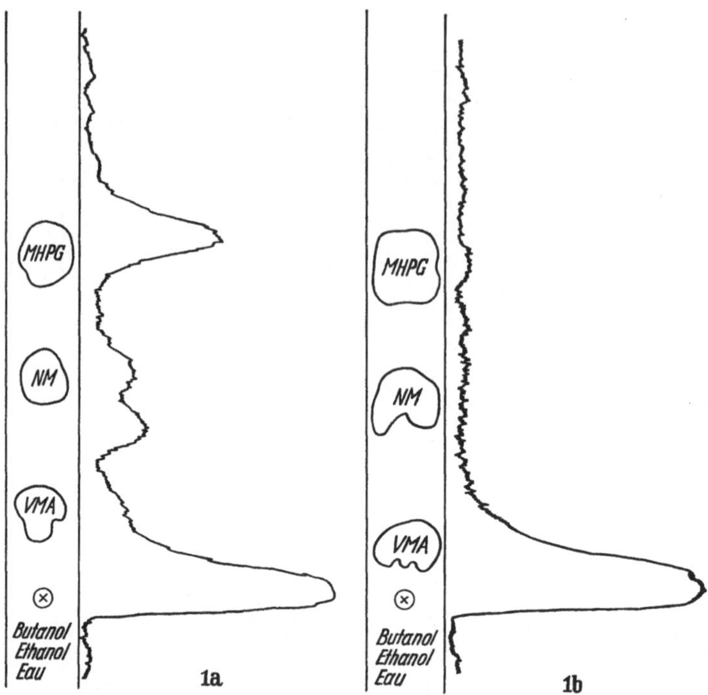

Fig. 4. Chromatogrammes des milieux de cultures 48 heures après l'addition de 7-³H noradrénaline (chlorhydrate). (a) Radiogramme du milieu avec une tumeur de la crête neurale; (b) radiogramme du milieu sans tumeur. Abréviations: VMA, acide 3-méthoxy-4-hydroxymandélique; NM; normetanéphrine, MHPG, 3-méthoxy-4-hydroxyphénylglycol

2° Il semble que le VMA lui-même n'est pas le produit formé dans la tumeur mais résulte du métabolisme du MHPG.

Enfin, l'étude cinétique du VMA a montré que d'autres métabolites sont très importants et grâce à cette étude nous avons obtenu une meilleure compréhension du métabolisme de la noradrénaline dans les tumeurs de la crête neurale.

Il reste la question de savoir ensuite s'il existe une différence entre les isomères D- et L- en égard à la vitesse des échanges entre les divers compartiments et à l'excrétion. Encore beaucoup de travail sera nécessaire avant que nous puissions fournir une réponse à cette question.

Remerciements

Ce travail a été réalisé en collaboration avec Mlle le Dr Schweisguth (Chef du Service de Pédiatrie) et le Pr Agrégé Bohuon (Chef de l'Unité de Biologie Chimique et de Biologie Expérimentale) où je travaille depuis 1 an, le Dr Barski et son assistant le Dr Belheradek, qui nous ont fourni les cultures de neuroblastomes. En effet, nous avons étudié le surnageant des cellules tumorales en cultures après avoir introduit de la noradrénaline marquée dans ces dernières.

Je remercie très vivement toutes ces personnes ainsi que les chirurgiens (Dr PEL-LERIN et Dr LEMOINE) les infirmières de l'Institut Gustave Roussy et des Hôpitaux des Enfants Malades et Bretonneau, enfin je remercie plus spécialement Mme GUENON et Mlle LAMBERT, laborantines de l'Unite de Biologie Expérimentale de l'I.G.R.

Bibliographie

[1] AMSTRONG, M. D., A. McMILLAN, and K. N. F. SHAW: 3-methoxy-4-hydroxy-D-mandelic acid, a urinary metabolite of norepinephrine. Biochem. Biophys. Acta 25, 422 (1957).
[2] —, K. N. F. SHAW, and P. E. WALL: J. biol. Chem. 218, 293 (1956).
[3] LA BROSSE, E. H.: Catecholamine metabolism: Studies on turnover rates of 3-methoxy-4-hydroxymandelic Acid. Abstr. Sixth Internat. Congr. Biochem. No. V-A-11, p. 391 (1964).
[4] —, J. BELEHRADEK, G. BARSKI, C. BOHUON, and O. SCHWEISGUTH: Metabolic activity of neural crest tumors in tissue culture. Nature 203, 194 (1964).
[5] — Observations non publié.

Résumé

L'auteurs résume des recherches effectuées en utilisant le VMA marqué par le tritium (H³VMA) de très haute activité spécifique. Avec ce produit on peut ainsi déterminer l'importance des divers compartiments du VMA. Ces résultats ainsi que ceux observés en culture de tissu indiquent que, au moins dans certains cas, la tumeur excrête non pas le VMA mais le MHPG.

Summary

In this paper, an approach to the knowledge of secretion of neuroblastomas is described. With the use of ³HVMA with high specific activity, it is possible to know the importance of pools of VMA. These results with others in tissue culture indicate that the tumor excrete not VMA but MHPG.

From the Bernhard Baron Memorial Research Laboratories and Institute of Obstetrics and Gynaecology, Queen Charlotte's Maternity Hospital, London, England

4-Hydroxy-3-Methoxyphenylglycol and other Compounds in Neuroblastoma

By

M. SANDLER and C. R. J. RUTHVEN

With 2 Figures

Tumours belonging to the neuroblastoma group have been the focus of intense biochemical interest since ISAACS et al. [1] first noted their ability to secrete catechol-amines. It soon became obvious that most, if not all, of these tumours are primarily dopamine-secreting [2—6], with a high urinary output of such metabolites as homo-vanillic acid [7—11] and dihydroxyphenylacetic acid [7, 12]. However, they also give rise to a large overproduction of noradrenaline, as indicated by an excess of β-hydroxylated metabolites in the urine of affected patients [13—15].

The excretion of noradrenaline metabolites, especially when age is taken into account, tends to be higher in neuroblastoma and ganglioneuroma [4, 9, 10, 16—18], than in phaeochromocytoma [19—23]. Paradoxically, only a minority of subjects with neuroblastoma have a marked elevation in blood pressure [4, 12, 17] whilst phaeochromocytoma, as is well known, gives rise to a severe degree of hypertension [24—28]. This would appear to be one of the central problems involved in the study of catecholamine-secreting tumours.

One possible explanation for this puzzling finding is that only a relatively small proportion of free noradrenaline enters the circulation of the patient with neuroblastoma, metabolic degradation occurring either in the tumour tissue itself or at peripheral sites, from local breakdown of noradrenaline arising from intracellular β-hydroxylation of circulating dopamine.

Recent experimental work using isotopically labelled catecholamines indicates that noradrenaline is likely to have a quantitatively different degradation pathway if it is liberated into the bloodstream than if it is directly metabolised by enzyme systems in close proximity to its binding sites. In the first case, O-methylation predominates whilst in the second, oxidative deamination also assumes a major role [29—32].

It therefore occurred to us that an assessment of the ratios of O-methylated to oxidatively deaminated urinary metabolites in both types of tumour might shed some light on this problem. Although techniques were well established for the assay of the major O-methylated metabolites, normetadrenaline (NMA) and metadrena-

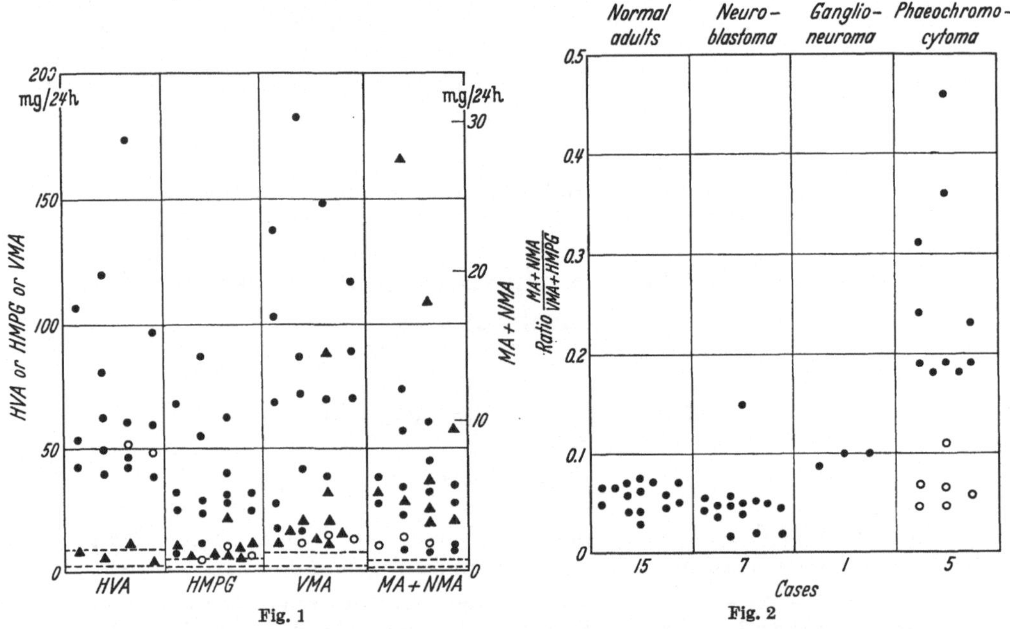

Fig. 1

Fig. 2

Fig. 1. Urinary excretion of catecholamine metabolites in neuroblastoma (●), ganglioneuroma (○) and phaeochromocytoma (▲). The normal ranges are indicated between the broken lines

Fig. 2. The ratio of urinary MA plus NMA to VMA plus HMPG in normal adults and various tumours (●), including post-operative phaeochromocytoma (○)

line (MA) [33—36], laboratory procedures for measurement of deaminated products were largely confined to 4-hydroxy-3-methoxymandelic acid (VMA) [37—39]; no method other than paper chromatographic was available for assay of 4-hydroxy-3-methoxyphenylglycol (HMPG) which quantitatively is almost as important as VMA. It was therefore necessary to devise one.

Based on previous work by AXELROD et al. [40], LA BROSSE et al. [41] and KOPIN et al. [42], a technique involving separation by ion-exchange chromatography and oxidation to vanillin was evolved which not only gave an accurate indication of output in normal individuals but also provided a more exact procedure, based on that of PISANO [33], for measurement of the metadrenalines [43]. In eighteen normal young adults, HMPG excretion ranged between 1.9 and 5.2 mg./24 hr. (mean 3.0 ± 0.8 [S.D.] mg./24 hr.), whilst 0.3—0.8 mg. total metadrenalines were excreted in 24 hr. (0.5 ± 0.14 [S.D.] mg./24 hr.).

Table 1

Reference	Diagnosis	Number of subjects	Number of specimens	Ratios			
				$\frac{MA + NMA}{VMA}$		$\frac{MA + NMA}{VMA + HMPG}$	
				Range	Mean	Range	Mean
This series	Normal controls: Adult Children (Age's 5¾ to 11 yr.)	16 4	18 4	0.041 to 0.11 0.068 to 0.127	0.083 0.106	0.028 to 0.076 0.054 to 0.10	0.055 0.083
This series	Neuroblastoma and Ganglioneuroma	10	21	0.024 to 0.21	0.086	0.017 to 0.148	0.060
GJESSING [8]	Neuroblastoma and Ganglioneuroma	8	8	0.013 to 0.15	0.077	0.012 to 0.118	0.055
This series	Phaeochromocytoma	10	16	0.26 to 0.56	0.37	0.18 to 0.44	0.27
KELLEHER et al. [23]	Phaeochromocytoma	18	29	0.17 to 2.1	0.50	—	—

Using this procedure, and that of PISANO et al. [39] for VMA, a clear-cut difference was observed between patients with neuroblastoma-ganglioneuroma and those with phaeochromocytoma In the latter group there was a greater proportion of metadrenalines, although, in general, there was a much smaller absolute excretion of VMA and HMPG than in neuroblastoma (Fig. 1). Patients with neuroblastoma-ganglioneuroma showed a mean ratio of O-methylated to oxidatively deaminated metabolites of less than 0.1, similar to that of normal adults, whilst those with phaeochromocytoma (whose ratio dropped to normal levels after operation) showed a mean value greater than 0.2 (Fig. 2).

The range and mean values for these ratios in controls, neuroblastoma and phaeochromocytoma are given in Table 1, together with ratio's calculated from the relevant metabolite output in cases of neuroblastoma or ganglioneuroma reported by GJESSING [8] and in patients with phaeochromocytoma reported by KELLEHER et al. [23]. In spite of methodological differences in measuring the respective catecholamine derivatives, the ratios derived from the results of Gjessing [8] and Kelleher *et al.* [23] support our finding that the excretion pattern of noradrenaline and adrenaline metabolites in neuroblastoma-ganglioneuroma was distinct from that observed in phaeochromocytoma. The ages of the children with neuroblastoma, in both our series and Gjessing's, ranged from one week to $9^{1}/_{2}$ years. Ratios in a small group of normal children were similar to those of normal adults and neuroblastoma-ganglioneuroma; there was no overlap with any of the values calculated for patients with phaeochromocytoma.

Neuroblastoma urines contained large amounts of homovanillic acid (Fig. 1.) the major metabolite of dopamine [10] confirming that all were dopamine secretors.

The proportionately large increase of metadrenalines in phaeochromocytoma has been stressed previously as a diagnostic aid [23, 44], although the dramatic difference in metabolite ratios described here between this tumour and neuroblastoma has not been noted before.

Although some caution must be observed in interpreting results based on a change in ratios [45, 46], we feel that our findings can best be explained by analogy with the experimental animal work of Kopin's group mentioned previously [29—32]; the high ratio in phaeochromocytoma, associated with marked hypertension, suggests a direct liberation of noradrenaline into the bloodstream, whereas the normal ratio found in the relatively normotensive neuroblastoma patient points to breakdown of noradrenaline before it has gained direct access to the circulation.

Summary

In this paper, the authors studies the ratios of O methylated to oxidatively deaminated urinary metabolites in neuroblastomas and pheochromocytomas. They described a new method for determination of MHPG. The normal secretion is ranged between 1.9 and 5.2 mg./24 H. The ratio of metabolites of NE gives a new argument for the breakdown of the amine in situ in neuroblastomas.

References

[1] Isaacs, H., H. Medalie, and W. M. Politzer: Brit. med. J. 1, 401 (1959).
[2] Smellie, J. M., and M. Sandler: Proc. roy. Soc. Med. 54, 327 (1961).
[3] Sandler, M., and C. R. J. Ruthven: Proc. V. int. Congr. Biochem., Moscow, p. 362 (1961).
[4] Voorhess, M. L., and L. I. Gardner: J. clin. Endocr. 22, 126 (1962).
[5] Rosenstein, B. J., and K. Engelmann: J. Pediat. 63, 217 (1963).
[6] Käser, H., M. Bettex, and W. von Studnitz: Arch. Dis. Childh. 39, 168 (1964).
[7] Studnitz, W. von: Scand. J. clin. Lab. Invest. 12, Suppl. 48, p. 53 (1960).
[8] Gjessing, L. R.: Scand. J. clin. Lab. Invest. 15, 463 (1963).
[9] Käser, H., O. Schweisguth, K. Sellei, and G. A. Spengler: Helv. med. Acta 30, 628 (1963).
[10] Ruthven, C. R. J., and M. Sandler: Analyt. Biochem. 8, 282 (1964).
[11] Brett, E. M., T. E. Oppé, C. R. J. Ruthven, and M. Sandler: Arch. Dis. Childh. 39, 403 (1964).
[12] Sourkes, T. L., R. L. Denton, G. F. Murphy, B. Chavez, and S. St. Cyr: Pediatrics 31, 660 (1963).
[13] Studnitz, W. von: Klin. Wschr. 40, 163 (1962).
[14] Voorhess, M. L., L. K. Pickett, and L. I. Gardner: Amer. J. Surg. 106, 33 (1963).
[15] Williams, C. M., and M. Greer: J. Amer. med. Ass. 183, 836 (1963).
[16] Bell, M.: In: "The Clinical Chemistry of Monoamines". Ed. by Varley, H., and Gowenlock, A. H., Amsterdam: Elsevier 1963, p. 82.
[17] Studnitz, W. von, H. Käser, and A. Sjoerdsma: New Engl. J. Med. 269, 232 (1963).

[18] Käser, H., M. Bettex, and W. von Studnitz: Arch. Dis. Childh. 39, 168 (1964).
[19] Gitlow, S. E., M. Mendlowitz, E. Kruk, and S. Khassis: Circulation Res. 9, 746 (1961).
[20] Jacobs, S. L., C. Sobel, and R. J. Henry: J. clin. Endocr. 21, 315 (1961).
[21] Sandler, M., and C. R. J. Ruthven: Biochem. J. 80, 78 (1961).
[22] Crout, J. R., and A. Sjoerdsma: J. clin. Invest. 43, 94 (1964).
[23] Kelleher, J., G. Walters, R. Robinson, and P. Smith: J. clin. Path. 17, 399 (1964).
[24] Hume, D. M.: Amer. J. Surg. 99, 458 (1960).
[25] Gitlow, S. E., M. Mendlowitz, and R. I. Wolf: J. Mount Sinai Hosp. 28, 159 (1961).
[26] Stackpole, R. H., M. M. Melicow, and A. C. Uson: J. Pediat. 63, 315 (1963).
[27] Gifford, R. W., W. F. Kvale, F. T. Maher, G. M. Roth, and J. T. Priestley: Mayo Clin. Proc. 39, 281 (1964).
[28] Hermann, H., and R. Mornex: "Human Tumours secreting Catecholamines." Oxford: Pergamon Press 1964.
[29] Kopin, I. J., G. Hertting, and E. Gordon: J. Pharmacol. 138, 34 (1962).
[30] —, and E. Gordon: J. Pharmacol. 138, 351 (1962).
[31] Kopin, I. J., and E. Gordon: J. Pharmacol. 140, 207 (1963).
[32] — Pharmacol. Rev. 16, 179 (1964).
[33] Pisano, J. J.: Clin. chim. Acta 5, 406 (1960).
[34] Studnitz, W. von: Clin. chim. Acta 5, 447 (1960).
[35] Smith, E. R., and H. Weil-Malherbe: J. Lab. clin. Med. 60, 212 (1962).
[36] Taniguchi, K., Y. Kakimoto, and M. D. Armstrong: J. Lab. clin. Med. 64, 469 (1964).
[37] Studnitz, W. von, and A. Hanson: Scand. J. clin. Lab. Invest. 11, 101 (1959).
[38] Weise, V. K., R. K. McDonald et E. H. La Brosse: Clin. chim. Acta 6, 79 (1961).
[39] Pisano, J. J., J. R. Crout et D. Abraham: Clin. chim. Acta 7, 285 (1962).
[40] Axelrod, J., I. J. Kopin et J. D. Mann: Biochim. biophys. Acta 36, 576 (1959).
[41] La Brosse, E. H., J. D. Mann, and S. S. Kety: J. Psychiat. Res. 1, 68 (1961).
[42] Kopin, I. J., J. Axelrod, and E. Gordon: J. biol. Chem. 236, 2109 (1961).
[43] Ruthven, C. R. J., and M. Sandler: Submitted for publication.
[44] Crout, J. R., J. J. Pisano, and A. Sjoerdsma: Amer. Heart J. 61, 375 (1961).
[45] Brunjes, S.: New Engl. J. Med. 271, 743 (1964).
[46] Wurtman, R. J.: New Engl. J. Med. 271, 743 (1964).

Institut Gustave Roussy, Villejuif (Seine), France

Intérêt pour le Pédiatre des Dosages d'Acide Vanillylmandélique dans les Urines

Par

O. Schweisguth

Avec 7 Figures

Nous avons entrepris, depuis 1960, l'étude de l'excrétion de l'acide vanillylmandélique (VMA) dans les urines de malades atteints de neuroblastome. Jusqu'à la fin de 1962, les dosages on été aimablement effectués à Berne par le Dr Kaser, les urines recueillies lui étant expédiées après acidification par 1 cc d'acide perchlorique 4 N par 100 ml d'urines. Depuis 1963, les examens ont été faits au Laboratoire de Biologie Experimentale de l'Institut Gustave Roussy à Villejuif, sous la Direction du Professeur Ag. Bohuon, et depuis juillet 1963 avec la précieuse collaboration de E. La Brosse. Les urines ont alors été recueillies sans acidification mais gardées à $+4°$ avant d'être portées au laboratoire.

Dans la mesure du possible, nous avons étudié les urines de tous les cas de neuroblastomes qui nous étaient adressés, et nous remercions tous les médecins et chirurgiens qui nous ont permis de réunir un abondant matériel. Chaque fois que cela a été possible, des dosages multiples ont été effectués au cours de l'évolution et du traitement de la maladie.

72 malades ont eu un ou plusieurs dosages, soit dès le début de leur maladie, soit lors d'une poussée évolutive plus tardive, mais pendant une *phase active* de leur tumeur. De plus, 17 enfants, antérieurement traités pour un neuroblastome, ont eu un ou plusieurs examens de contrôle, alors qu'ils étaient apparemment stabilisés depuis une ou plusieurs années. Enfin, de nombreux enfants atteints de tumeurs non sympathiques ou d'autres affections ont été examinés et ont servi de témoins.

Le diagnostic de neuroblastome a été établi sur des arguments cliniques et radiologiques. Il a été contrôlé dans 50 cas par l'histologie et 19 cas par la cytologie dans les formes en évolution; pour deux cas métastatiques avec forte secrétion, nous n'avons pas de document histologique; le dernier cas est le cas IGR 63—5684. Les cas «stabilisés» ont été vérifiés histologiquement, sauf deux où le diagnostic repose sur la clinique et la radiologie (tumeurs rétro-péritonéales refoulant le rein et calcifiées, une fois avec métastases hépatiques cliniques).

1. Résultats

La Fig. 1 résume les résultats des dosages dans nos cas; sur cette figure, en cas de dosages multiples, nous n'avons indiqué que le premier dosage effectué. Les résultats ont été exprimés en μg par mg de créatinine; malgré les réserves qu'appelle ce mode d'expression, il semble le moins mauvais.

Une cause d'erreur doit être notée: l'infection des urines par un colibacille a provoqué chez une de nos malades habituellement très secrétante, la disparition totale du VMA; de plus, l'adjonction de VMA dans ces urines infectées n'a pas permis de retrouver en chromatographie une réaction positive et il faut admette que le VMA a été détruit en présence de colibacilles. Un seul dosage négatif ne donne donc pas une assurance absolue qu'il n'existe pas d'excrétion urinaire anormale.

Nos témoins, sauf deux pour lesquels nous n'avons pas d'explication satisfaisante, ont tous excrété mois de 20 μg, et la plupart moins de 10 μg/mg de créatinine (Fig. 1).

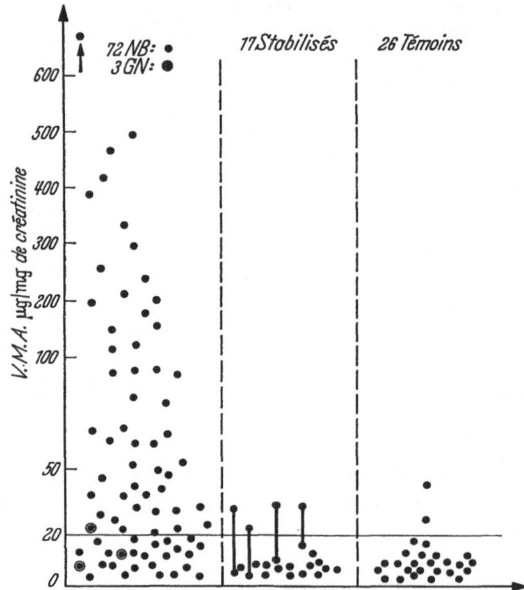

Fig. 1. Répartition des taux d'excrétion de VMA dans 72 neuroblastomes, 3 ganglióneuromes, 17 neuroblastomes stabilisées, et 26 temoins. Les dosages successifs, dans 4 neuroblastomes stabilisés sont reliés par un trait

Les enfants ayant eu un neuroblastome stabilisé avaient en règle des valeurs normales, sauf 4 ayant eu des formes inopérables et traités par RX et éventuellement chimiothérapie; chez les 4, les dosages ultérieurs ont montré le retour à un taux normal, mais ceci longtemps après la disparition de tout signe clinique (Fig. 1).

Chez les enfants porteurs de neuroblastomes évolutifs, les résultats se sont étagés entre 0 et 1000 μg/mg de créatinine: la Fig. 2 montre la répartition des cas selon l'importance de l'excrétion: dans 21 cas un dosage inférieur à 20 μg ne peut être considéré comme pathologique. Ceci ne veut pas forcément dire qu'il s'agit d'une forme non secrétante, car l'acide homovanillique peut alors être isolément augmenté de façon significative (VMA 6 μg et HVA 355 μg dans un cas).

Dans les ³/₄ des cas cependant, une élévation franche du taux de VMA dans les urines peut apporter au clinicien un argument de poids en faveur du sympathome avant toute histologie, et permettre ainsi un diagnostic:

— entre tumeur polaire supérieure du rein et sympathome, dans les formes rétropéritonéales,

— entre sympathome et tumeur dys-embryoplasique dans les formes média-stinales,

— entre lymphosarcome et sym-pathome dans les formes ganglionnaires cervicales,

— entre sarcome d'Ewing et sym-pathome dans certaines formes à géné-ralisation osseuse.

Ceci est d'autant plus important que, dans ces divers diagnostics, la cytologie et même parfois l'histologie, ne per-mettent pas toujours de trancher avec certitude.

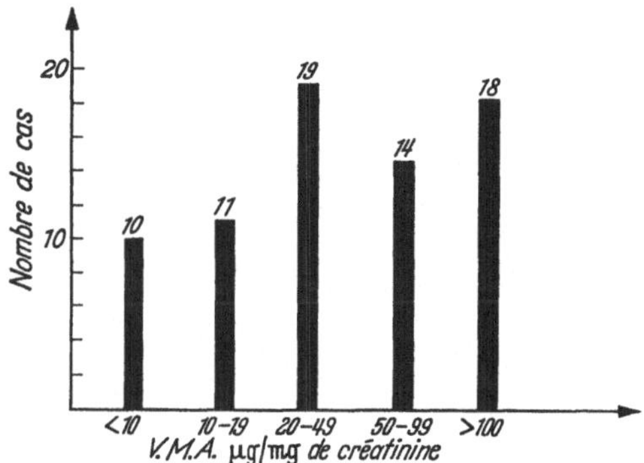

Fig. 2. Répartition de 72 cas de neuroblastomes en évo-lution par rapport à l'excrétion de VMA

2. Histologie

Nous n'avons pu trouver aucune différence dans l'aspect histologique des diverses tumeurs: ni le degré de différenciation, ni l'importance de la nécrose et des calcifi-cations, ni la richesse en embolies vasculaires ou lymphatiques ne permettent de pré-voir l'importance de la secrétion tumorale.

3. Tension artérielle

Il serait séduisant d'imaginer un parallèlisme entre la valeur de la pression artérielle et la secrétion tumorale, dans une tumeur produisant des catécholamines; mais les choses ne sont pas simples.

La prise de la pression artérielle est difficilement précise chez le petit enfant: dans un cas médiastinal où elle avait été évaluée soigneusement à 105/60 avant inter-vention chez un nourrisson de 10 mois, la prise de tension per-opératoire par cathéter intra-fémoral, avec enregistrement continu, la montrait à 120/80, sous anesthésie, avant l'ouverture du thorax.

D'autre part, la grande majorité des neuroblastomes siègent dans la région juxta-rénale, comprimant par eux-mêmes ou par des adénopathies satellites un pédicule rénal, et parfois les deux. Une hypertension peut être alors liée à ce mécanisme: un nourrisson de 7 mois avait une tension artérielle à 115/60 avec une secrétion de VMA normale, mais une grosse tumeur englobant le pédicule rénal droit.

La Fig. 3 montre dans 29 cas les rapports de la tension arterielle et de l'exretion de VMA, au même moment de l'évolution; dans l'ensemble, les pressions maxima sont plutôt élevées, mais il n'y a aucun parallèlisme entre les deux données; ceci irait bien avec la notion que la noradrénaline synthétisée par la tumeur est détruite avant d'être libérée dans l'organisme.

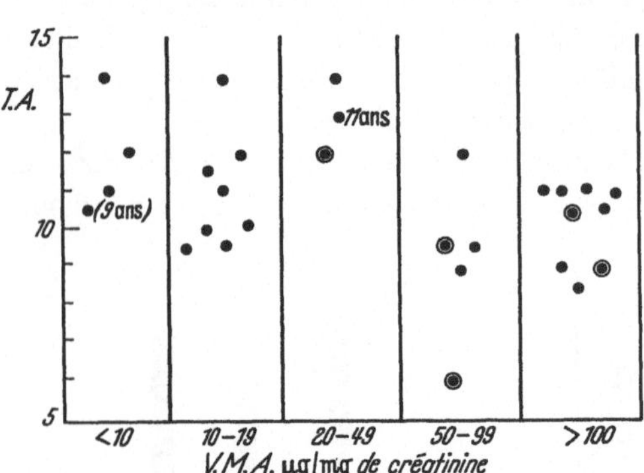

Fig. 3. Tension artérielle et excrétion de VMA dans 29 cas de neuroblastomes. 27 enfants sur 29 ont moins de 5 ans, et généralement moins de 3 ans.   ⊙ localisations mediastinales

4. Caractères cliniques

Age: 6 de nos malades seulement avaient plus de 5 ans; une forte élimination de VMA parait plus constante chez les nourrisons de moins d'un an.

a) Localisation

Les formes abdominales sont de beaucoup les plus fréquentes, mais ne se distinguent pas des autres par leur secrétion.

Nous avons observé deux cas de neuroblastome de la région narinaire, typiques histologiquement, mortels après métastases ganglionnaires disséminées, et hépatiques dans un cas, — qui n'étaient pas secrétants.

b) Généralisation

Malgré des différences probables dans le nombre de cellules secrétantes, nous n'avons pas trouvé de différences notables entre les formes loco-régionales, et les formes largement métastatiques. Cependant, 5 cas de métastases hépatiques massives, cliniquement isolées, du petit nourrisson, montraient une excrétion massive de VMA.

5. Traitements

Chaque fois que cela nous a été possible, nous avons essayé de comparer les modifications biochimiques aux effets cliniques des traitements.

a) Chirurgie (Fig. 4)

Dans 15 cas, nous avons eu des dosages avant et après chirurgie. 7 fois, après ablation complète de la tumeur, le taux de VMA est revenu à la normale, mais dans un cas, au bout de 3 semaines seulement.

Une fois, l'ablation d'une très grosse tumeur rétro-péritonéale, laissant en place des ganglions envahis adhérents à l'aorte, a fait tomber le dosage de VMA de 400 à 30 μg, semblant montrer l'importance de la masse de tissu tumoral; mais dans un autre cas, une simple biopsie (B — Fig. 4) limitée d'une volumineuse tumeur avec envahissement ganglionnaire a aussi fait baisser l'excrétion de 100 à 20 μg...

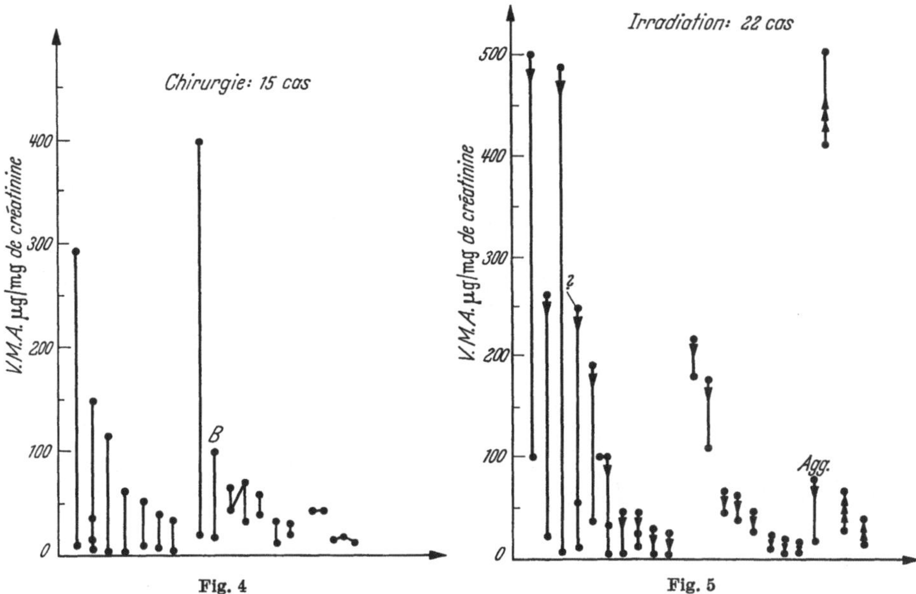

Fig. 4. Evolution de l'excrétion de VMA après traitement chirurgical. Les dosages concernant le même malade sont reliés par un trait. B: biopsie

Fig. 5. Evolution de l'excrétion de VMA après radiothérapie. Les dosages concernant le même malade sont reliés par un trait. Agg: aggravation clinique

Dans 4 cas, la baisse du VMA a été limitée malgré une chirurgie macroscopiquement complète (parfois, le taux reviendra à la normale après irradiation post-opératoire). Deux fois, le taux de VMA n'a pas changé après ablation de la tumeur. Le volume tumoral ne conditionne donc pas à lui seul l'importance de l'excrétion.

b) Radiothérapie (Fig. 5)

L'action de l'irradiation est plus difficile à juger, surtout dans les formes métastatiques où l'ensemble du tissu tumoral n'est pas toujours irradié. L'évolution biochimique et clinique est en général parallèle. Dans 10 cas, la baisse du taux de VMA a été très spectaculaire, mais plus ou moins rapide: deux fois, la diminution clinique de la tumeur et la baisse du VMA n'ont débuté qu'après la fin de l'irradiation pour aboutir à la normale après plusieurs semaines.

Dans trois cas, l'inefficacité des RX s'est accompagnée d'une augmentation de l'excrétion de VMA. Une fois (Fig. 5), dans l'irradiation de métastases généralisées, le VMA a baissé sous traitement, tandis que l'enfant mourait du développement de ses lésions.

Enfin, nous avons eu trois malades dont le VMA était revenu à la normale sous l'irradiation, qui ont présenté une nouvelle poussée tumorale métastatique, sans remontée de la secrétion tumorale: on peut penser alors que cette nouvelle multiplication cellulaire s'est faite aux dépens de cellules biologiquement modifiées par le traitement antérieur.

c) Chimiothérapie (Fig. 6)

Nous avons habituellement utilisé la cyclophosphamide, en injections intra-musculaires, à des doses de 7 à 10 mg par Kg de poids et par jour, pendant 7 à 10 jours, selon la tolérance sanguine. Parallèlement à un bon effet clinique, nous avons observé des baisses remarquables du taux de VMA, alors que dans deux cas sans amélioration clinique, l'excrétion a augmenté.

Dans l'ensemble, l'évolution des dosages de VMA au cours des traitements des sympathomes montre beaucoup d'anomalies déconcertantes: mais il faut penser que la tumeur n'est pas obligatoirement un tout homogène: elle peut comprendre des zones inactives parce que nécrotiques, et il est aussi possible que toutes les cellules qui la composent ne soient pas également douées des mêmes propriétés biologiques, ou de la même sensibilité aux traitements; et l'excrétion urinaire de VMA ne représente que la résultante de toute la tumeur. Aussi est-il nécessaire de chercher d'autres méthodes d'étude et d'analyse de la biochimie de ces tumeurs.

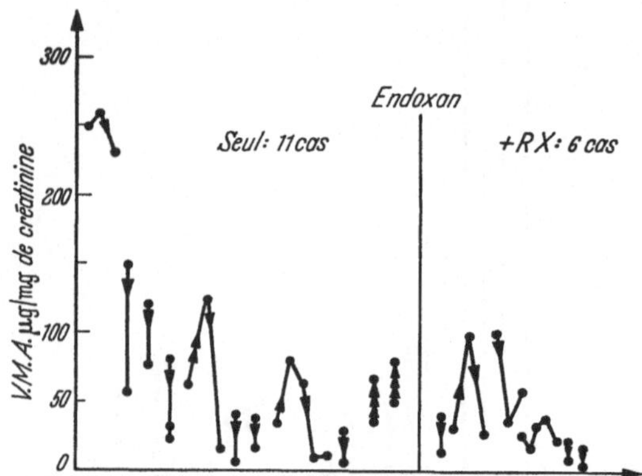

Fig. 6. Evolution de l'excrétion de VMA sous chimiothérapie (cyclophosphamide) seule ou associée à la radiothérapie. Les dosages concernant le même malade sont reliés par un trait

Une observation est rapportée à titre d'exemple. CO . . . . . Paola (IGR 63.5684) 3 mois (Fig. 7), est hospitalisée en octobre 1963, avec une petite tumeur non calcifiée supra-rénale gauche, et un foie métastatique qui remplit tout l'abdomen. Son excrétion de VMA est si forte, et son état clinique si précaire qu'on renonce à toute biopsie. Elle reçoit en urgence en quelques jours 250 rads sur son foie avec une amélioration clinique et biologique nette. On décide ensuite d'essayer l'efficacité de la vitamine B 12 dans ce cas relativement favorable, et l'enfant reçoit 1000 µg tous les deux jours pendant plus de 3 mois. Pendant ce temps, son foie ne se modifie pas; son poids oscille avec des épisodes de diarrhée avec déshydratation, et on aperçoit à deux reprises de petites ecchymoses orbitaires. Aussi fait-on fin février une cure de 300 mg au total d'Endoxan; quelques jours après la fin de ce traitement, le foie ne s'est toujours pas modifié, et on retrouve une ecchymose orbitaire, et un discret feuilleté périosté d'un fémur. On irradie alors le foie à la dose de 1100 rads tumeur, et le fémur à la dose de 1000 rads. L'amélioration clinique est alors spectraculaire: le foie régresse rapidement tandis que la courbe de poids devient pour la première fois régulièrement ascendante. Parallèlement, le VMA s'abaisse régulièrement pour revenir à la normale en avril 64. L'enfant a été revue le 31 Mai 64, ayant pris encore un kilog, avec un foie débordant à peine d'un travers de doigt, et un VMA à moins de 5 µg/mg de créatinine. Elle est en parfait état en juillet 65.

Chez elle, il semble bien que les RX aient été l'agent thérapeutique le plus efficace.

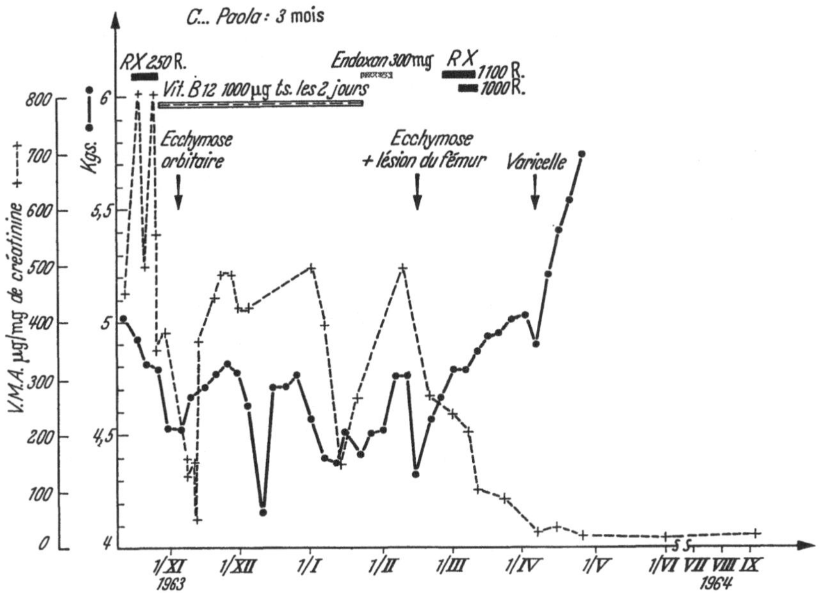

Fig. 7. Evolution clinique et biochimique d'un cas de neuroblastome avec métastases hépatiques
——×—: dosages de VMA
—●—: poids

Résumé

Dans cette communication l'auteur étudie en relation avec les observations cliniques, les dosages de VMA dans les urines. Cet examen permet de suivre l'action d'une chimiothérapie, d'une radiothérapie, ou d'une intervention chirurgicale. Cet article met bien en évidence la nécessité pour le clinicien de faire effectuer ces dosages dans le cas d'un neuroblastome.

Summary

The clinical observations are correlated with the estimations of VMA. Furthermore, the evolution of VMA concentration in urine is followed in relation with chemotherapy, radiotherapy and surgery. The great importance of VMA estimation and the usefulness for the clinician are underlined.

Bibliographie

KÄSER, H.: Schweiz. med. Wschr. 91, 586—589 (1961).
KONTRAS, S. B.: Cancer Chemother. Rep. 16, 443—453 (1962).
MASON, G. A., J. HART-MERCIER, E. J. MILLAR, L. B. STRANG, and N. A. WYNNE: Lancet 1957, 322—325.
SOURKES, T. L., R. L. DENTON, G. F. MURPHY, B. CHAVEZ, and S. ST CYR: Pediatrics 31, 660—668 (1963).
STICKLER, G. B., and E. V. FLOCK: Cancer Chemother. Rep. 16, 439—442 (1962).
VOORHESS, M. L., and L. I. GARDNER: Pediatrics 30, 241—247 (1962).
—, and J. P. WHALEN: Radiology 83, 92—97 (1964).
WILLIAMS, C. M.: J. Amer. med. Ass. 183, 836—840 (1963).

M. GOLDSTEIN

Neurochemistry Laboratories, Department of Psychiatry and Neurology
New York University Medical Center, New York 16, New York

Enzymes Controlling the Biosynthesis of Catecholamines [1]

By

M. GOLDSTEIN [2]

With 1 Figure

Isotopic studies have confirmed the proposed major pathway for the biogenesis of norepinephrine and epinephrine [1]. The main steps in this pathway are the oxidation of phenylalanine to tyrosine followed by hydroxylation to 3,4-dihydroxy-phenylalanine, decarboxylation to dopamine, β-hydroxylation of dopamine to nor-epinephrine and N-methylation of norepinephrine to epinephrine. In addition, some

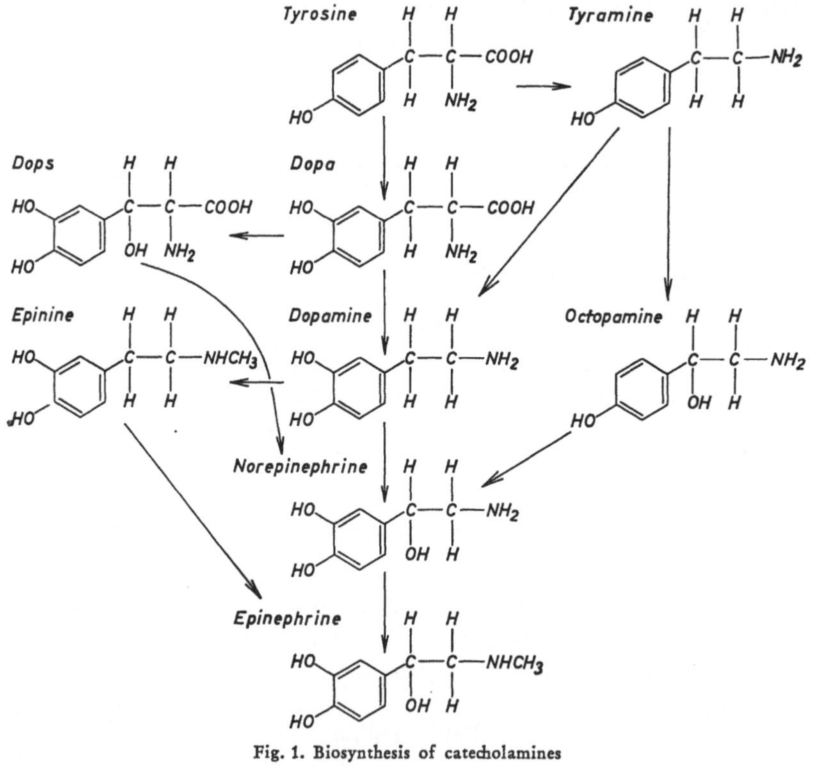

Fig. 1. Biosynthesis of catecholamines

other minor pathways were postulated (Fig. 1). Although evidence for the main pathway was well established a decade ago, only recently, all the enzymes which control the biosynthesis of catecholamines were isolated and their properties charac-terized. Based on these studies it is now possible to investigate the activities of these

[1] Dedicated to Professor I. Äbelin, Bern, Switzerland.

[2] Research Career Development Awardee of the United States Public Health Service, Grant. No. 5-K3-MH-14,1918-04 and also supported by the Grant of USPHS No. MH-02717-06.

enzymes in tumors such as neuroblastoma and pheochromocytoma, and to determine which of those enzymes is responsible for the overproduction of catecholamines.

The purpose of this paper is first, to review the properties of the enzymes which catalyze the biosynthesis of catecholamines, and second, to give a preliminary report on the comparative activities of tyrosine hydroxylase and dopamine-β-hydroxylase in a neuroblastoma tumor and in a pheochromocytoma tumor.

1. Phenylalanine Hydroxylase

The enzyme phenylalanine hydroxylase catalyzes the hydroxylation of phenylalanine to tyrosine. This enzyme was isolated from a soluble fraction of the liver [2]. It was also shown that the aromatic hydroxylation requires as a co-factor a tetrahydropteridine derivative which is oxidized to a "dihydropteridine" during the formation of tyrosine from phenylalanine. The inactivation of the enzyme after incubation with versene or cysteine can be reversed with ferrous ion, but it is not known whether a metal is essential for the enzymatic activity. The enzyme utilizes O_2 and not H_2O as the source of oxygen in the conversion of phenylalanine to tyrosine.

2 Tyrosine Hydroxylase

It was recently demonstrated that brain, adrenal medulla and sympathetically innervated tissue contain a specific hydroxylase that catalyzes the conversion of L-tyrosine to DOPA [3]. Soluble preparations of the enzyme were obtained from beef adrenal medulla [4]. It was also shown that the purified enzyme requires as a co-factor, tetrahydropteridine derivatives. Fe^{++} ions also stimulate the enzymatic activity. Some aromatic L-amino acids such as α-methyl dopa, α-methyl tyrosine, phenylalanine were found to be effective inhibitors [4]. Preliminary studies have shown that 3-iodo-tyrosine and other thyroid hormone derivatives are inhibitors of tyrosine hydroyxlase.

The finding that tyrosine hydroxylase requires the same tetrahydropteridines as phenylalanine hydroxylase shows the importance of folic acid derivatives in catecholamine biosynthesis.

3. DOPA Decarboxylase

An enzyme purified from the soluble fraction of guinea pig kidney decarboxylates a variety of aromatic amino acids. The enzyme has been named aromatic L-amino acid decarboxylase [5]. This enzyme requires as a co-factor pyridoxal phosphate. The enzyme also decarboxylates its own competitive inhibitors, namely α-methyl amino acids [6]. The hydrazino analogue of α-methyl dopa is an effective inhibitor but it does not readily cross the blood-brain barrier [7].

4. Dopamine-β-Hydroxylase

The enzyme which catalyzes the conversion of dopamine to norepinephrine has been purified from chromaffin granules of the adrenal medulla [8]. A further purification of the enzyme was recently obtained by chromatography of the enzyme on DEAE cellulose column [9]. The enzyme requires ascorbic acid and fumaric acid as co-factors [8]. Without ascorbic acid a catechol compound can serve as an alternative reducing agent [10]. The enzyme is non-specific and accepts a variety of substrates including phenylethylamine, tyramine, epinine, 3-methoxy dopamine, amphetamine and other phenylethyl and phenylpropylamines [10, 11, 12]. The configuration of the α-carbon is decisive as to whether β-hydroxylation will occur. Only d-α-methyl phenylethylamines are substrates for the hydroxylating enzyme [13].

The enzymatic β-hdroxylation of phenylethylamines and phenylpropylamines may be physiological significance and may influence the biosynthesis and metabolism of catecholamines. It is possible that the synthesis of norepinephrine *in vivo* may occur via octopamine. It was indeed shown that after administration of tyramine-C^{14}, norepinephrine-^{14}C was excreted in the urine [14]. Tyramine and octopamine are not substrates of tyrosine hydroxylase, and the conversion of tyramine and octopamine to norepinephrine is catalyzed by an enzyme isolated from liver homogenates [15]. In sympathetically innervated tissues such as the heart norepinephrine is formed from tyrosine, but not from tyramine or octopamine [16]. Since epinine has been shown to be a substrate for the hydroxylating enzyme [11], its conversion to epinephrine may represent a new route for the biosynthesis of this hormone. Although 3-methoxy dopamine is a weak substrate for the hydroxylating enzyme [11], it is possible that some of the 3-methoxy norepinephrine formed *in vivo* is derived from 3-methoxy dopamine and not only from norepinephrine.

The lack of substrate specificity of the hydroxylating enzyme suggest that many dopamine analogues which are not substrates may act as inhibitors. Indeed, it was shown that dopamine analogues, such as adrenalone or arterenone are effective inhibitors [17]. Benzylhydrazine and benzyloxylamines are also active competitive inhibitors [18].

The inhibition of dopamine-β-hydroxylase by a variety of chelating agents suggests that dopamine-β-hydroxylase is a metalloenzyme [19, 20]. The effective inhibition of dopamine-β-hydroxylase by strong Cu binding agents such as tropolones and diethyldithiocarbamate and the fact that the metal is tightly bound to the protein suggest that dopamine-β-hydroxylase is a Cu enzyme. Disulfiram was found to be a very potent inhibitor of dopamine-β-hydroxylase *in vivo* and *in vitro* [21]. It was also shown that inhibition of dopamine-β-hydroxylase by disulfiram is effective enough to make β-hydroxylation the rate-limiting step in norepinephrine synthesis [22, 23].

5. Phenylethanol Amine-N-Methyl Transferase

The supernatant fraction of beef adrenal medullary homogenates has been shown to form epinephrine-^{14}C from noradrenaline, ATP and methionine methyl-^{14}C [24]. The enzymatic reaction is similar to the methyl transferring systems described by CANTONI [25]. The enzyme was purified from the soluble fraction of adrenal medulla and was shown to be specific for phenylethanol amines [26]. A non-specific N-methylase from the soluble fraction of rabbit lung with S-adenosyl-methionine as a co-factor methylates a variety of biogenic amines such as catecholamines, serotonin, tryptamine, histamine and mescaline [27].

6. Tyrosine Hydroxylase and Dopamine-β-Hydroxylase Activities in a Neuroblastoma and Pheochromocytoma Tumor

Various kinds of tumors which arise from the adrenal medulla and from the sympathetic nervous system have been found to contain large quantities of catecholamines. Pheochromocytoma tumors have been reported to contain high levels of norepinephrine and epinephrine associated with an increased excretion rate of VMA, the major metabolite of these catecholamines [28, 29, 30]. Neuroblastoma and ganglioneuroblastoma tumors have also been reported to contain high levels of catecholamines associated with an increased excretion rate of VMA and HVA, the latter being the major excretion product of dopamine [31, 32]. It was therefore of special interest to investigate the activities of the enzymes which control the biosynthesis of catecholamines in pheochromocytoma and neuroblastoma tumors. The rate-limiting steps in norepinephrine biosynthesis are most likely catalzyed by tyrosine hydroxylase and dopamine-β-hydroxylase [22, 23, 4]. The activities of these two enzymes were assayed in these tumors.

Tyrosine hydroxylase activity was determined in a crude tissue extract prepared from the tumor according to the procedure of NAGATSU et al. [3]. Dopamine-β-hydroxylase was isolated from the tumor particles in the same manner as described for the isolation of the hydroxylating enzyme from the normal adrenal medulla particles [8]. Aliquots of the solubilized enzyme preparation were incubated as previously described [11]. The results of these studies are summarized in Table 1.

Table 1. *Tyrosine hydroxylase and dopamine-β-hydroxylase activity in a neuroblastoma and pheochromocytoma tumor*

	Dopa formed from tyrosine mμmoles/g/hr.	Norepinephrine formed from dopamine μmoles/g/10 min.
Neuroblastoma Tumor . .	30—35	0.00
Pheochromocytoma Tumor	0.00	5.5

It is evident that tyrosine hydroxylase activity was detected in the neuroblastoma tumor but not in the pheochromocytoma. On the other hand, very high dopamine-β-hydroxylase activity was detected in the pheochromocytoma tumor but not in the neuroblastoma tumor. Tyrosine hydroxylase activity in neuroblastoma tumors has been reported by others at this meeting [33]. It is of interest that we have not detected tyrosine hydroxylase activity in the pheochromocytoma tumor. It is conceivable that the high levels of norepinephrine in pheochromocytoma are only due to the high levels of dopamine-β-hydroxylase and that tyrosine hydroxylase is not present in such tumors. However, it is also possible that tyrosine hydroxylase activity was inhibited in this incubation mixture by norepinephrine which was present in the crude tumor homogenate [4]. Some further studies on dopamine-β-hydroxylase and tyrosine hydroxylase activities in neuroblastoma and pheochromocytoma tumors are in progress and will be reported elsewhere.

Summary

The enzymes which control the biosynthesis of catecholamines are reviewed, specially, the properties and the relation of these enzymes with nervous tumors. A comparison of tyrosine hydroxylase and dopamine hydroxylase activities in pheochromocytoma and neuroblastoma is made. The author find a striking difference between these two enzymes.

References

[1] BLASCHKO, H.: In M. Harington (Editor): Hypotensive drugs. London: Pergamon Press 1956, p. 23.

[2] KAUFMAN, S.: J. Biol. Chem. 234, 2677 (1959).

[3] NAGATSU, R., M. LEVITT, and S. UDENFRIEND: Biochem. Biophys. Res. Commun. 14, 543 (1964).

[4] — — — J. Biol. Chem. 239, 2910 (1964).

[5] LOVENBERG, W., H. WEISSBACH, and S. UDENFRIEND: J. Biol. Chem. 237, 89 (1962).

[6] OATES, J. A., L. GILLESPIE, S. UDENFRIEND, and A. SJOERDSMA: Science (Washington) 131, 1890 (1960).

[7] PORTER, C. C., L. S. WATSON, D. C. TITUS, J. A. TOTARO, and S. S. BYER: Biochem. Pharmacol. 11, 1067 (1962).

[8] LEVIN, E. Y., B. LEVENBERG, and S. KAUFMAN: J. Biol. Chem. 235, 2080 (1960).

[9] GOLDSTEIN, M.: To be published.

[10] LEVIN, E. Y., and S. KAUFMAN: J. Biol. Chem. 236, 2043 (1961).

[11] GOLDSTEIN, M., and J. F. CONTRERA: J. Biol. Chem. 237, 1898 (1962).

[12] CREVELING, C. R.: Ph. D. Thesis, Georgetown University, Washington D. C. 1962.

[13] GOLDSTEIN, M., M. R. MAC KEREGHAN, and E. LAUBER: Biochim. biophys. Acta 89, 191 (1964).

[14] CREVELING, C. R., M. LEVITT, and S. UDENFRIEND: Life Sci. 1, 523 (1962).

[15] LEMBERGER, L.: Fed. Proc. 23, 2758 (1964).

[16] MUSACCHIO, J. M., and M. GOLDSTEIN: Biochem. Pharmacol. 12, 1061 (1963).

[17] GOLDSTEIN, M., and J. M. MUSACCHIO: Biochem. Pharmacol. 11, 809 (1962).

[18] CREVELING, C. R., J. B. VAN DER SCHOOT, and S. UDENFRIEND: Biochem. Biophys. Res. Commun. 8, 215 (1962).

[19] GOLDSTEIN, M., E. LAUBER, and M. R. MCKEREGHAN: Biochem. Pharmacol. 13, 1103 (1964).

[20] GREEN, A. L.: Biochim. biophys. Acta 81, 394 (1964).

[21] GOLDSTEIN, M.: Life Sci. 3, 763 (1964).

[22] MUSACCHIO, J. M.: Life Sci. 3, 769 (1964).

[23] GOLDSTEIN, M.: To be published.

[24] KIRSHNER, N.: Biochim. biophys. Acta 24, 658 (1957).

[25] CANTONI, G.: Phosphorus Metabolism, vol. I p. 641. Ed. by W. D. McElroy and B. Glass. Baltimore: Johns Hopkins Press 1951.

[26] AXELROD, J.: J. Biol. Chem. 237, 1657 (1962).

[27] — J. Pharmacol. exp. Ther. 138, 28 (1962).

[28] ARMSTRONG, M. D., A. McMILLAN, and K. N. F. SHAW: Biochim. biophys. Acta 25, 422 (1957).

[29] KELLY, H. M.: Proc. Staff Meet. Mayo Clin. 11, 65 (1936).

[30] GOLDENBERG, M.: Amer. J. Med. 10, 627 (1951).

[31] VOORHESS, M. L., and L. I. GARDNER: J. clin. Endocr. 21, 321 (1961).

[32] STUDNITZ, W. VON: Lancet 1961, 215.

[33] — This meeting.

Conclusion

Ce Colloque devait avant tout permettre à un certain nombre de biochimistes spécialisés dans l'étude des catécholamines de faire le point des connaissances acquises depuis quelques années dans la biochimie des tumeurs des ébauches sympathiques. Le but souhaité était de permettre de se faire une idée aussi précise que possible sur la valeur de telle ou de telle détermination des catécholamines ou de leurs métabolites pour le diagnostic et la surveillance thérapeutique, et sur la physiopathologie de ces tumeurs.

Plusieurs communications ont définitivement montré que le dosage de l'acide vanylmandélique (VMA) est indispensable, et que jusqu' à présent aucune autre détermination n'est aussi intéressante pour le clinicien. Le dosage de cet acide phénolique peut se faire par différentes techniques, mais la plupart des participants pensent qu'en dehors de la méthode chromatographique décrite lors de la découverte du VMA par ARMSTRONG, la méthode de PISANO (ou une de ses nombreuses modifications publiées) est excellente. En dehors de ce dosage fondamental, les opinions sont aussi unanimes sur l'intérêt que présente le dosage de l'acide homovanillique ainsi que celui de la dopamine. Malheureusement, les techniques de dosage de ces deux molécules sont beaucoup plus complexes. Il n'en est pas moins vrai que dans certains neuroblastomes on n'observe pas d'augmentation de VMA dans les urines, alors que les teneurs de HVA et de dopamine sont anormalement élevées.

Seules des études longues et portant sur un nombre élevé de malades pourront nous dire l'importance de telle ou telle autre détermination (MHPG, dopa etc. . . .). Le nombre des cas encore classés comme non secrètants diminueront peut être au fur et à mesure que le spectre des dosages chimiques (dans le sang ou dans l'urine) pourra être élargi. Il reste encore un énorme travail de mises au point techniques, de confrontations des statistiques etc. . . . pour avoir une opinion même assez fragmentaire.

Dans un deuxième groupe de communications, des travaux de recherche on été abordés.

Il en ressort que les déterminations d'activité enzymatique dans les tumeurs obtenues chirurgicalement, ou après mise en culture de tissue, laissent espérer des classifications possibles, fort importantes peut être pour le diagnostic et la thérapeutique. En ce sens, le travail de GOLDSTEIN, s'il est confirmé sur un plus grand nombre de tumeurs, montre une différence fondamentale entre le phéochromocytome (présence de dopamine β hydroxylase, absence de tyrosine hydroxylase) et le tissu neuroblastique (absence de dopamine β hydroxylase et présence de tyrosine hydroxylase).

Le travail de LA BROSSE et coll. a montré également la présence et l'activité des enzymes catabolisant les catécholamines en culture de tissu et surtout a été le point de départ de travaux sur la secrétion de MHPG par le tissu neuroblastique. Ce travail explique en partie l'absence d'hypertension constatée dans la très grande majorité des neuroblastomes. D'autres communications ont apporté des éléments intéressants sur le métabolisme de la dopa, de la dopamine etc. . . .

En définitive, les possibilités de recherche offertes aux cancérologues par cette tumeur de l'enfant, sont larges, mais des limitations sont très souvent imposées par la difficulté de se procurer du sang ou les tissus en quantités suffisantes pour les études biochimiques. Il est d'autre part gênant de ne disposer chez l'animal d'aucune tumeur, même greffée, présentant des analogies de secrétion. Une autre difficulté est due à la rareté de molécules standards nécessaire aux recherches.

En ce sens, une discussion fort intéressante a pris naissance lors de ce colloque, à la suite de la communication du Professeur GARDENT, sur les meilleures façons d'aborder ces problèmes de chimie de synthèse. Des échanges de molécules rares entre les divers participants ont déjà été nombreux et fort utiles dans différentes expérimentations.

En conclusion, tous les participants ont souhaité se réunir dans un délai d'environ 2 ans pour confronter les progrès réalisés dans ce domaine. Il est certain que des réunions groupant un nombre relativement restreint de spécialistes sur un sujet très précis, sont à l'heure actuelle la meilleure façon de faire progresser nos connaissances. Certains points nouveaux discutés lors de ce premier colloque comme l'étude en microscopie électronique, la culture de tissus nerveux, le facteur de croissance des nerfs (NGF), les renouvellements du VMA radio-actif etc. . . . seront certainement largement traités dans le futur.

Pr Agrégé C. BOHUON   Docteur Odile SCHWEISGUTH
Institut G. Roussy   Institut G. Roussy

Gesamtherstellung: Konrad Triltsch, Graphischer Großbetrieb, Würzburg